# SpringerBriefs in Materials

The SpringerBriefs Series in Materials presents highly relevant, concise mono-graphs on a wide range of topics covering fundamental advances and new applications in the field. Areas of interest include topical information on innovative, structural and functional materials and composites as well as fundamental principles, physical properties, materials theory and design. SpringerBriefs present succinct summaries of cutting-edge research and practical applications across a wide spectrum of fields. Featuring compact volumes of 50 to 125 pages, the series covers a range of content from professional to academic. Typical topics might include:

- A timely report of state-of-the-art analytical techniques
- A bridge between new research results, as published in journal articles, and a contextual literature review
- A snapshot of a hot or emerging topic
- An in-depth case study or clinical example
- A presentation of core concepts that students must understand in order to make independent contributions

Briefs are characterized by fast, global electronic dissemination, standard publish-ing contracts, standardized manuscript preparation and formatting guidelines, and expedited production schedules.

More information about this series at http://www.springer.com/series/10111

Pratima Bajpai

# Carbon Fibre from Lignin

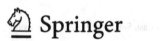
Springer

Pratima Bajpai
Pulp and Paper, Consultants
Kanpur
India

ISSN 2192-1091          ISSN 2192-1105   (electronic)
SpringerBriefs in Materials
ISBN 978-981-10-4228-7   ISBN 978-981-10-4229-4   (eBook)
DOI 10.1007/978-981-10-4229-4

Library of Congress Control Number: 2017933854

Printed on acid-free paper

This Springer imprint is published by Springer Nature
The registered company is Springer Nature Singapore Pte Ltd.
The registered company address is: 152 Beach Road, #21-01/04 Gateway East, Singapore 189721, Singapore

# Preface

This book presents detailed information on the production and properties of carbon fibres derived from lignin precursors. Focusing on future directions of the carbon fibre industry, it also introduces a novel process for obtaining high-purity lignin, a key aspect in the manufacture of high-quality carbon fibre. Carbon fibre is currently the most preferred lightweight manufacturing material and is rapidly becoming the material of choice for manufacturers around the world. Although more than 80% of commercial carbon fibre is estimated to use PAN (polyacrylonitrile) as a precursor, carbon fibre manufactured from PAN is expensive and therefore its application is limited to high-performance structural materials. After cellulose, lignin is the second most abundant natural biopolymer and offers a carbon-rich, renewable resource. As a byproduct of the pulp and paper industry and the production of cellulosic ethanol, lignin is available at low cost. It is an economically attractive alternative to the use of PAN for the production of carbon fibres, as highlighted in this book. The information presented will be of interest to all those involved in the investigation of carbon fibre materials, carbon fibre manufacturers and carbon fibre users.

Kanpur, India                                                                                       Pratima Bajpai

# Acknowledgements

Some excerpts taken from Bajpai Pratima (2013). "Update on Carbon Fiber" Smithers Rapra, UK with kind permission from Smithers Rapra © 2013, Smithers Rapra Technology Ltd., UK.

Some excerpts taken from Baker DA, Rials TG. (2013). Recent advances in low-cost carbon fiber manufacture from lignin. J Appl Polym Sci;130:713–28, DOI: 10.1002/APP.39273 with kind permission from John Wiley & Sons Inc., USA, Copyright © 2013, John Wiley & Sons, Inc.

# Contents

1   **General Background and Introduction** .......................... 1
     References........................................................... 8

2   **Lignin**.............................................................. 11
     References........................................................... 14

3   **Carbon Fibre**...................................................... 17
     References........................................................... 22

4   **Carbon Fibre Market** ............................................. 25
     References........................................................... 28

5   **Lignin as a Raw Material for Carbon Fibre**.................... 29
     References........................................................... 31

6   **Recovery of Lignin** ............................................... 33
     6.1   Kraft Lignin ................................................. 34
     6.2   Steam Explosion Lignin ................................... 38
     6.3   Organosolv Lignin ........................................ 39
     References........................................................... 41

7   **Lignin as a Precursor for Carbon Fibre Production** ........... 43
     7.1   Production from Different Types of Lignin................. 44
     References........................................................... 57

8   **Conversion of Lignin to Carbon Fibre** ......................... 63
     8.1   Wet Spinning .............................................. 63
     8.2   Dry Spinning .............................................. 63
     8.3   Melt Spinning.............................................. 64
     References........................................................... 66

9   **Future Directions**................................................. 69
     References........................................................... 70

**Index** ................................................................. 73

# List of Figures

Figure 1.1  Chemical structure of hardwood and softwood lignin.
Based on Nimz (1974), Gargulak and Lebo (1999) . . . . . . . . .    5

Figure 1.2  Common Lignin monomers . . . . . . . . . . . . . . . . . . . . . . . . .    6

Figure 1.3  Lignin products and their market value.
Based on Frank (2013) . . . . . . . . . . . . . . . . . . . . . . . . . . . . .    7

Figure 2.1  Common bond in lignin. The most important bond
in lignin is the β-O-4. Based on Henriksson (2007) . . . . . . . .   13

Figure 3.1  Process for manufacture of carbon fibre from PAN . . . . . . . . .   21

Figure 3.2  Process for manufacture of carbon fibre from Pitch . . . . . . . . .   22

Figure 6.1  Kraft pulping and chemical recovery process.
Based on Smook (1992) . . . . . . . . . . . . . . . . . . . . . . . . . . . .   34

Figure 6.2  Organosolv fractionation method. Based on Baker
and Rials (2013), Bozell et al. (2011) . . . . . . . . . . . . . . . . . .   40

Figure 8.1  Production steps involved in the production of carbon
fibre from lignin. Based on Baker and Rials (2013),
Chen (2014) . . . . . . . . . . . . . . . . . . . . . . . . . . . . . . . . . . . .   64

# List of Tables

Table 1.1  Use of carbon fibre in different industries. . . . . . . . . . . . . . . . . .  2
Table 1.2  Repartition of the production costs of PAN-based
          carbon fibres. . . . . . . . . . . . . . . . . . . . . . . . . . . . . . . . . . . . . . .  3
Table 1.3  Lignin linkages in softwood and hardwood . . . . . . . . . . . . . . .  6
Table 2.1  Composition of biomass. . . . . . . . . . . . . . . . . . . . . . . . . . . . . . . .  12
Table 2.2  Types of lignin and their use . . . . . . . . . . . . . . . . . . . . . . . . . .  14
Table 3.1  Properties of carbon fibres and its comparison
          with other materials . . . . . . . . . . . . . . . . . . . . . . . . . . . . . . . . .  18
Table 3.2  Classification of carbon fibre . . . . . . . . . . . . . . . . . . . . . . . . . .  19
Table 3.3  Properties of carbon fibre . . . . . . . . . . . . . . . . . . . . . . . . . . . . .  20
Table 3.4  Disadvantages of carbon fibre. . . . . . . . . . . . . . . . . . . . . . . . . .  20
Table 4.1  Global carbon fibre demand in thousand tones . . . . . . . . . . . . .  26
Table 4.2  Carbon fibre capacity by manufacturer
          in 1000 tonnes (2014). . . . . . . . . . . . . . . . . . . . . . . . . . . . . . . .  27
Table 6.1  Suppliers of lignin . . . . . . . . . . . . . . . . . . . . . . . . . . . . . . . . . . .  37
Table 6.2  Advantages of steam explosion processes. . . . . . . . . . . . . . . . .  38
Table 7.1  Properties of carbon fibre from different hardwood lignin. . . . . .  46
Table 7.2  Physical properties of carbon fibres from blends
          of lignin-PEO. . . . . . . . . . . . . . . . . . . . . . . . . . . . . . . . . . . . . . .  53
Table 7.3  Physical properties of carbon fibres from blends of lignin . . . . .  53
Table 7.4  Properties of carbon fibre from different types of lignin . . . . . . .  57

# Chapter 1
# General Background and Introduction

**Abstract** Of the carbon fibres produced today, over 90% originate from the oil-based synthetic polymer polyacrylonitrile, which is expensive to produce. Exploiting lignin as a precursor for carbon fibre adds high economic value to lignin and encourages further development in lignin extraction technology. General background and introduction on lignin as a potential precursor material for producing carbon fibre is presented in this chapter.

**Keywords** Lignin · Biopolymer · Renewable polymer · Carbon fibre · Polyacrylonitrile · High strength material

Carbon fibres are high-strength materials having a micro graphite crystal structure. These fibres contain at least 90% carbon and are obtained from pyrolysis of appropriate fibres under controlled conditions and are used in the manufacture of advanced composite materials (Figueiredo et al. 1990; Chung 1994; Watt 1985; Donnet and Bansal 1990). These are classified by the source materials from which they are derived, such as polyacrylonitrile (PAN), rayon and pitch. In the late 1950s, Roger Bacon produced the first high-performance carbon fibres (Bacon 1960). After that rayon-based carbon fibres which were made up of rayon strands processed by carbonation were developed. In the beginning of 1960s, PAN-based carbon fibres were developed by Akio Shindo (Nakamura et al. 2009). Leonard Singer developed pitch-based carbon fibres in 1970 (Singer and Lewis 1978). To manufacture PAN-based carbon fibres, PAN is processed into fibrous shape by spinning, and then oxidation, carbonization and surface treatment are conducted.

In industrial applications, carbon fibres were rarely used and were generally used to reinforce or add function to composite materials with a base material of plastic, ceramic, metal and so on. Carbon fibre commonly has superior mechanical properties including high specific strength and high specific modulus. They also possess characteristics such as low density, low thermal expansion, heat resistance and chemical stability. Carbon fibre is lightweight, has high strength, shows good flexibility and fatigue-resistance. These properties result from the orientation of carbon atoms along the fibre axis during the manufacturing process. In addition,

© The Author(s) 2017                                                                                     1
P. Bajpai, *Carbon Fibre from Lignin*, SpringerBriefs in Materials,
DOI 10.1007/978-981-10-4229-4_1

various kinds of carbon fibres having differing mechanical performance or fibre morphology are developed.

Globally, carbon fibre is in rapidly growing demand for the different industries (Figueiredo et al. 1990; Chung 1994; Watt 1985; Donnet and Bansal 1990; Minus and Kumar 2005, 2007; Fitzer et al. 1989; Hajduk 2005; Huang 2009; Saito et al. 2011; Barnes et al. 2007; Soutis 2005; Ogawa 2000; Nolan 2008; van der Woude et al. 2006; Fuchs et al. 2008; Zhang and Shen 2002; Aoki et al. 2009; Tran et al. 2009; Olenic et al. 2009; Baughman et al. 2002; Thostensona et al. 2001; Bajpai 2013; Kubo and Kadla 2005a, b) as shown in Table 1.1. Carbon nanofibres open possibilities for applications in regenerative medicine and cancer treatment too (Barnes et al. 2007; Soutis 2005; Ogawa 2000; Nolan 2008; van der Woude et al. 2006; Fuchs et al. 2008. Zhang and Shen 2002; Aoki et al. 2009; Tran et al. 2009; Olenic et al. 2009; Baughman et al. 2002; Thostensona et al. 2001).

In the recent years, carbon fibres have found wide applications in the aircraft commercial and civilian, recreational, industrial and transportation markets. Carbon fibres are used in composites with a lightweight matrix. Carbon fibre composites are best suited for those applications where stiffness, strength, lower weight and outstanding fatigue characteristics are the major requirements. They also can be used in the situation where high temperature, chemical inertness and high damping are required.

In the early 1960s successful commercial production of carbon fibre started, as the requirements of the aerospace industry particularly for military aircraft—for lightweight materials—became of greatest importance (Pimenta and Pinho 2011; Wood 2010). The application of composite materials to various parts of aircraft had started in the 1970s and facilitated weight reduction of aircraft (Soutis 2005). Currently, in some types of aircraft, carbon fibre composites comprise more than 10% of the total weight, and such applications shall increase in the future. Carbon fibre composites are also applied to rockets and satellites (Vignoles et al. 2010). In athletics, they are used in the production of fishing rods, golf club shafts, tennis rackets, and so on to achieve weight reduction or improvement of rigidity and durability (McHenry 1984; Ogawa 2000). In the medicine, utilizing its radiolucency, carbon fibre-reinforced plastics (CFRPs) are often used for X-ray devices X-ray device (Baidya et al. 2001). Also, CFRPs, which are lightweight and easy to handle, are employed for medical devices such as limb prostheses and wheel chairs (Nolan 2008; van der Woude et al. 2006). Moreover, carbon fibres are used in construction and civil engineering in materials for seismic strengthening or

| **Table 1.1** Use of carbon fibre in different industries | Aeronautics |
| --- | --- |
| | Automotive |
| | Marine |
| | Transportation |
| | Construction |
| | Electronics |
| | Wind energy |

construction. In the automobile industry, development of lightweight and fuel-efficient cars using CFRPs in the structural parts is aggressively progressing (Fuchs et al. 2008). Applications of carbon fibres are also studied in the energy fields, including fuel cell development and oil drilling, and in electronic devices such as personal computers and liquid crystal projectors.

The main drawback is the high production cost, which limits the supply, despite a growing demand. The principal processing steps for the production of carbon fibre typically include spinning, stabilization, carbonization and sometimes graphitization (Bahl et al. 1998). Today, the polyacrylonitrile (PAN) precursor represents approximately half of the production costs (Table 1.2) and the equipment is associated with approximately one-third of the production costs (Warren et al. 2009). Many industries are interested in carbon fibres as a new and lightweight material with the potential to replace, for example, the steel in cars and the glass fibres in blades in wind power stations. The main precursor for the manufacturing of carbon fibre today is PAN, which constitutes approximately 90% of all commercial carbon fibre produced (Forrest et al. 2001). The two other precursors are petroleum pitch and regenerated cellulose (rayon). Regarding the properties of precursors, lignin is most similar to petroleum pitch.

Lignin is an aromatic biopolymer. It is amorphous, highly branched polyphenolic macromolecule with a complex structure, and the material typically forms about 1/3 of the dry mass of woody materials. It can be used as a precursor for the production of carbon fibres. It is derived from wood and plants and has a significant potential cost advantage over even textile-grade PAN as a precursor material for production of low-cost carbon fibre (www1.eere.energy.gov/vehiclesandfuels/pdfs/lm/7_low-cost_carbon_fibre.p).

It has been projected that the low cost of lignin would save 37–49% in the final production cost of carbon fibre. Replacement of PAN by lignin for wind turbine blades has a triple pay-off: it uses renewable resources, optimizes energy and materials costs for the production of carbon fibres, and the fibres themselves will be used in wind turbines for producing renewable energy. But as lignin is a breakable biopolymer, it cannot be spun, stretched/aligned, and spooled into fibres without modification.

Lignin is a highly abundant biopolymeric material (second only to cellulose) and is currently obtained from chemical pulping of wood. But now several biomass refineries are coming on stream, therefore, the lignin byproduct from cellulosic fuel ethanol production will represent a valuable resource material for carbon fibre production. Works on lignin produced using Oganosolv pulping have already

| Table 1.2 Repartition of the production costs of PAN-based carbon fibres | Precursor | 51% |
| --- | --- | --- |
| | Labour | 10% |
| | Depreciation | 12% |
| | Utilities | 18% |
| | Others | 9% |

Based on Warren et al. (2009)

shown that such type of lignins can be readily melt-spinnable as isolated. These lignins are of a much higher purity level compared to lignins produced from the chemical pulping of wood for paper production.

Of the three major natural polymers that make up ordinary plants—cellulose, lignin and hemicellulose—lignin is one of the major components of all vascular plants and the only biomass constituent based on aromatic units. It is an amorphous, polyphenolic material. It arises from the enzymatic dehydrogenative polymerization of three phenyl propanoid monomers, namely sinapyl alcohol, coniferyl alcohol and p-coumaryl alcohol. The biosynthesis process leads to the generation of a complex, three-dimensional polymer. This lacks the ordered, repeating units found in other natural polymers like cellulose and proteins. Lignin plays a very important role in the carbon cycle. It sequesters atmospheric carbon into the living tissues of woody plants. It actually fills the spaces in the cell wall between hemicellulose, cellulose and pectin components of wood. It is covalently bound to hemicellulose and by that means crosslinks the different plant polysaccharides and impart mechanical strength to the cell wall and to the plant as a whole. The structure of lignin from beech hardwood is shown in Fig. 1.1. Lignin structure is known to be significantly changed when lignocellulosic material is treated under conditions intended to separate the lignin from the cellulose. Lignin provides structure to woody materials and is the component mainly responsible for imparting strength to wood against mechanical stress. The chemical and physical properties of lignin vary depending upon the following factors:

- Wood species
- Botanical origin
- Region from which the wood is harvested
- Process by which the lignin is isolated.

Lignin contains of coniferyl alcohol and sinapyl alcohol units in varying ratios. Softwood lignin mainly contains coniferyl alcohol (>90%) and a small proportion of p-coumaryl alcohol Fig. 1.2 shows structure of coniferyl alcohol, sinapyl alcohol and p-coumaryl alcohol. Table 1.3 shows lignin linkages in hardwood and softwood. It is very difficult to isolate lignin from wood without degradation. Therefore, the actual molecular weight of native lignin is a subject of debate. The weight average molecular weight of softwood lignin isolated by mechanical milling of the wood is about 20,000. Milled hardwood lignin shows significantly lower molecular weights. These and other differences between the chemistries of softwood and hardwood lignins affect the utilization of lignin for carbon fibre production.

On a dry-wood basis, lignin contains about 15–25 wt% of wood, compared to about 38–48 wt% cellulose depending on the species. The lignin content of switch grass is similar to that of lignin. Thus, wood and switch grass contain significant amounts of lignin—about half of the proportion of cellulose, which if used for the production of value-added products, such as carbon fibre, could effectively counterbalance the high cost of producing cellulosic ethanol from biomass. A relatively small proportion of the lignin would be used as a fuel in the ethanol production

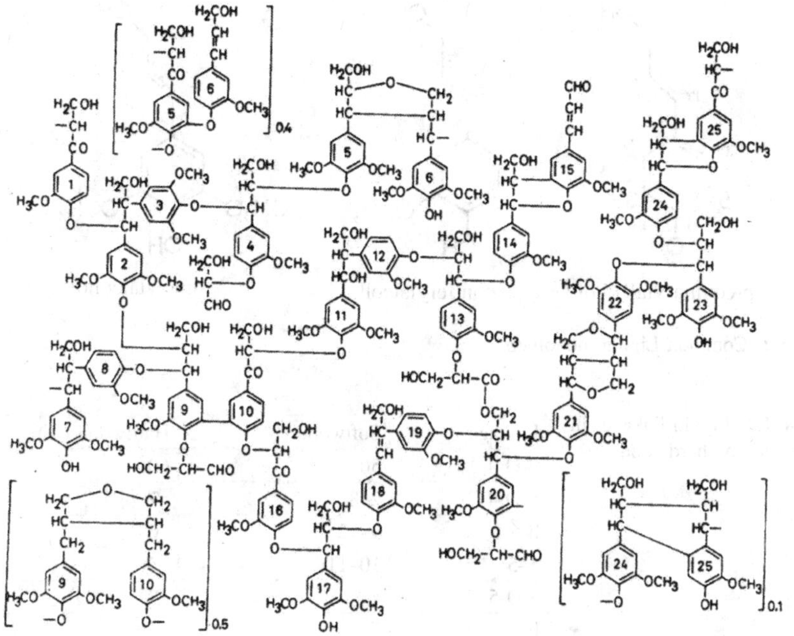

Hardwood lignin

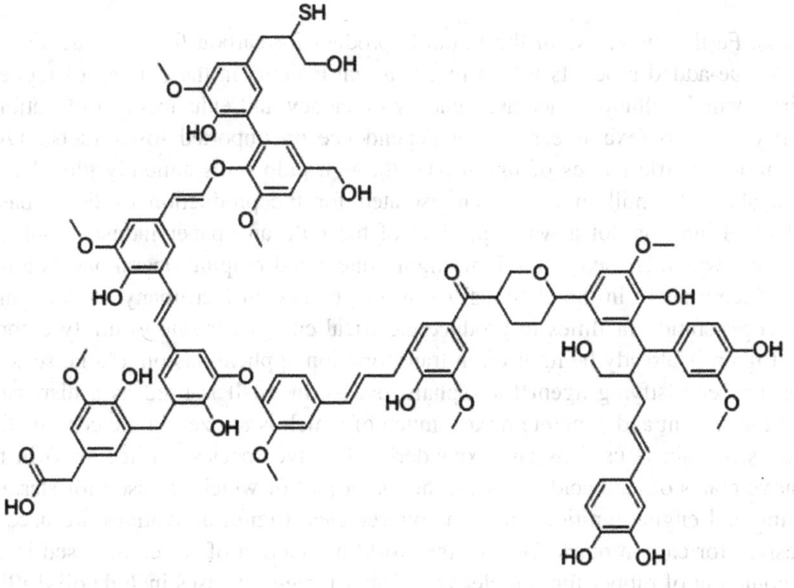

Softwood lignin

**Fig. 1.1** Chemical structure of hardwood and softwood lignin. Based on Nimz (1974), Gargulak and Lebo (1999)

p-coumarylalcohol        coniferylalcohol        sinapylalcohol

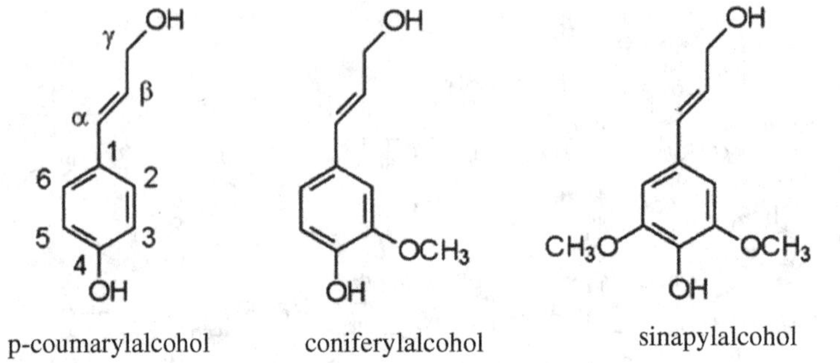

**Fig. 1.2** Common Lignin monomers

**Table 1.3** Lignin linkages in softwood and hardwood

| Linkage | Softwood (%) | Hardwood (%) |
|---------|--------------|--------------|
| β-O-4 | 50 | 60 |
| α-O-4 | 2–8 | 7 |
| β-5 | 9–12 | 6 |
| 5–5 | 10–11 | 5 |
| 4-O-5 | 4 | 7 |
| β-1 | 7 | 7 |
| β-β | 2 | 3 |

process. Furthermore, use of the lignin byproduct for carbon fibre production and other value-added products would result in real benefits in the context of reduced environmental pollution, increased energy efficiency and enhancement of national security interests (example reducing dependence on imported fossil fuels). Over 200 million metric tonnes of lignin pass through pulp mills annually globally, of which about 1.2 million tonnes are isolated for the production of lignin-based products. Lignin is not a waste product of the pulp and paper industry, but is a coproduct which is used in modern, highly integrated pulping operations as a fuel and reducing agent in the chemical recovery process. In fact, many modern mills have cogeneration facilities to produce electrical energy for sale to utility companies. Lignin is already being used in transportation applications on a large scale. It is used as emulsifying agent for asphalt road surfaces. It is used as a dispersing agent for cement and concrete mixes, much of which is utilized in the construction of roads. Lignin is used as an "expander" (of active species surface area) in the negative plates of lead-acid batteries, the major part of which are used for starting, lighting and engine ignition functions on vehicles. Lignin derivatives are used as adhesives for carbon black, 70% of the world production of which are used in the compounding of rubber for vehicle tyres. Other large-scale uses include oil-drilling muds; textile dye dispersants; sequestrants for micronutrients (agricultural and forestry uses); pesticide surfactants; binders for plywood, particle board, fibre glass insulation; animal feed and water treatment for boilers and cooling systems.

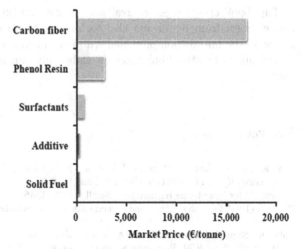

**Fig. 1.3** Lignin products and their market value. Based on Frank (2013)

The high carbon content in the lignin macromolecules makes it a potential candidate for carbon fibre production. Figure 1.3 shows lignin products and their market value.

Lignin is a carbon fibre precursor—an alternative to high-cost, petroleum-based precursors. However, a suitable precursor must also have fibre-forming ability by some of the spinning methods and withstand the following high temperature treatment. Thermoplasticity has been observed for lignins showing the possibility to use melt extrusion for fibre spinning, (Kadla et al. 2002a, b; Kubo et al. 2005) which is the preferred processing technique as it is less costly than wet spinning. The desire to produce lignin-based carbon fibre is not new. Already in the 1960s, a commercial carbon fibre was produced from lignin, called the Kayocarbon fibre. It was produced from lignosulphonates by dry spinning together with a plasticizer, polyvinyl alcohol (Otani et al. 1969). However, the production soon discontinued, while the PAN process was improved which enabled an ever since increasing market share for PAN-base carbon fibres (Lucintel 2011). Nowadays however, as mentioned earlier, the drive for finding alternatives for PAN-based carbon fibres is intensifying, because of increasing demands combined with higher raw material price and environmental awareness (Warren et al. 2009). Lignins from various wood pulping methods have been used in previous melt spinning studies, such as steam-explosion lignin, organosolv lignin, and kraft lignin. (Sudo and Shimizu 1992; Sudo et al. 1993; Kadla et al. 2002a; Kubo et al. 1997, 1998, 2005; Warren et al. 2009).

Carbon fibre application has the highest potential to add value to lignin at $21,700–$800,000 per tonne. This represents up to 2000 times more value recovered from lignin. The overall size of the carbon fibre precursor market is estimated to be $2.25 billion (50% of the carbon fibre tow market) (Chen 2014).

This book brings together available information on production, properties of carbon fibres from lignin and also focuses on future directions of carbon fibre industry. The information presented will be of interest to all those involved in the investigation of carbon fibre materials, carbon fibre manufacturers and carbon fibre users.

# References

Aoki K, Usui Y, Narita N, Ogiwara N, Iashigaki N, Nakamura K, Kato H, Sano K, Ogiwara N, Kametani K, Kim C, Taruta S, Kim YA, Endo M, Saito N (2009) A thin carbon-fibre web as a scaffold for bone-tissue regeneration. Small 5:1540–1546

Bacon R (1960) Growth structure and properties of graphite whiskers. J Appl Phys 1960(31):283–290

Bahl OP, Shen Z, Lavin JG, Ross RA (1998). Manufacture of carbon fibres. In: Donnet J-B, Wang TK, Peng JCM, Reboulliat S (eds) Carbon fibres, Marcel Dekker, New York, pp 1–83

Baidya KP, Ramakrishna S, Rahman M, Ritchie A (2001) Quantitative radiographic analysis of fibre reinforced polymer composites. J Biomater Appl 15:279–289

Bajpai P (2013) Update on carbon fiber. Smithers Rapra, UK

Barnes CP, Sell SA, Boland ED, Simpson DG, Bowlin GL (2007) Nanofibre technology: designing the next generation of tissue engineering scaffolds. Adv Drug Delivery Rev 59:1413–1433

Baughman RH, Zakhidov AA, De Heer WA (2002) Carbon nanotubes—the route toward applications. Science 297(5582):787–792

Chen MCW (2014) Commercial viability analysis of lignin based carbon fibre. Master thesis, Simon Fraser University

Chung DL (1994) Carbon Fibre Composites. Boston, MA, USA, Butterworth Heinemann, pp 3–65

Donnet JB, Bansal RC (1990) Carbon fibres, 2nd edn. Marcel Dekker, New York, pp 1–145

Figueiredo JL, Bernardo CA, Baker RTK, Huttinger KJ (eds) (1990) Carbon fibres filaments and composites. Kluwer Academic, Dordrecht, pp 327–336

Fitzer E, Edie DD, Johnson DJ (1989) In: Figueiredo JL, Bernardo CA, Baker, RTK, Huttinger KJ (eds) Carbon fibres filaments and composites, 1st edn. Springer, New York, pp 3–41, 43–72, 119–146

Frank K (2013). Lignin based carbon nanofibre. www.carbonfibreworkshop.com/wp-content/uploads/2013/08/2-Frank-Ko-UBC.pdf

Forrest A, Pierce, J, Jones W, Hwu, E (2001). Low-cost carbon fibres. Synergy. Retrieved August 18, from http://virtual.clemson.edu/caah/synergy/ISSUE-1LCCF.htm

Fuchs ERH, Field FR, Roth R, Kirchain RE (2008) Strategic materials selection in the automobile body: economic opportunities for polymer composite design. Compos Sci Technol 68:1989–2002

Gargulak JD, Lebo E (1999) Commercial use of lignin-based materials, volume 742 of ACS symposium series. American Chemical Society, pp 304–320

Hajduk F (2005). Carbon fibres overview. Global outlook for carbon fibres 2005. Intertech Conferences. San Diego, CA, 11–13 October 2005

Huang X (2009) Fabrication and properties of carbon fibres. Materials 2:2369–2403. doi:10.3390/ma2042369

Kubo S, Kadla JF (2005a) Kraft/lignin/poly(ethylene oxide) blends: effect of lignin structure on miscibility and hydrogen bonding. J Appl Polym Sci 98:1437–1444

Kubo S, Kadla JF (2005b) Lignin-based carbon fibres: effect of synthetic polymer blending on fibre properties. J Polym Environ 13(2):97–105

Kubo S, Ishikawa N, Uraki Y, Sano Y (1997) Preparation of lignin fibres from softwood acetic acid lignin: by atmospheric acetic acid pulping. Mokuzai Gakkaishi 43:655–662

Kubo S, Uraki Y, Sano Y (1998) Preparation of carbon fibres from softwood lignin by atmospheric acetic acid pulping. Carbon 36(7–8):1119–1124

Kadla JF, Kubo S, Gilbert RD, Venditti RA, Compere AL, Griffith WL (2002a) Lignin-based carbon fibres for composite fibre applications. Carbon 40(15):2913–2920

Kadla JF, Kubo, S, Gilbert, RD, Venditti, RA (2002b) In: Hu TQ (ed) Lignin-based carbon fibres, chemical modification, properties, and usage of lignin. Kluwer Academic/Plenum, New York, pp 121–137

Kubo S, Gilbert, RD, Kadla JF (2005). Lignin-based polymer blends and biocomposite materials. In: Natural fibres, biopolymers, and biocomposites, pp 671–697

Lucintel Market Report (2011) Growth opportunities in the global carbon fibre market: 2011–2016. Lucintel, Las Colinas

McHenry CR (1984) Exercise and sports equipment: some ergonomics aspects. Appl Ergon 15:259–279

Minus ML, Kumar S (2005) The processing, properties and structure of carbon fibres. JOM 57 (2):52–58

Minus ML, Kumar S (2007), Carbon fibre Kirk-Othmer encyclopedia of chemical technology, vol 26, pp 729–749

Nakamura O, Ohana T, Tazawa M, Yokota S, Shnoda W, Nakamura O, Itoh J (2009) Study on the PAN carbonfibre-innovation for modeling a successful R&D management. Synthesiology 2:154–164

Nimz HH (1974) Beech lignin—proposal of a constitutional scheme. Angew Chem Int Ed Engl 13:313–321

Nolan L (2008) Carbon fibre prostheses and running in amputees: a review. J Foot Ankle Surg 14:125–129

Ogawa H (2000) Architectural application of carbon fibres: development of new carbon fibre reinforced glulam. Carbon 38:211–226

Olenic L, Mihailescu G, Puneanu S, Lupu D, Biris AR, Margineanu P, Garabagiu, S (2009) Investigation of carbon nanofibres as support for bioactive substances. J Mat Sci Mat Med 20 (1):177–183. ISSN 0957-4530

Otani S, Fukuoka Y, Igarashi B, Sakaki K (1969). Method for producing carbonized lignin fibre. (Nippon Kayaku Kk), US Pat. 3461082

Pimenta S, Pinho ST (2011) Recycling carbon fibre reinforced polymers for structural applications: technology review and market outlook. Waste Manag 31:378–392

Saito N, Aoki K, Usui Y, Shimizu M, Hara K, Narita N, Ogihara N, Nakamura K, Ishigaki N, Kato H, Haniu H, Taruta S, Kim YA, Endo M (2011) Application of carbon fibres to biomaterials: a new era of nano-level control of carbon fibres after 30-years of development. Chem Soc Rev 40(7):3824–3834

Singer LS, Lewis IC (1978) ESR study of the kinetics of carbonization. Carbon 16:417–423

Soutis C (2005) Fibre reinforced composites in aircraft construction. Prog Aerosp Sci 2005 (41):143–151

Sudo K, Shimizu K (1992) A new carbon fibre from lignin. J Appl Polym Sci 44(1):127–134

Sudo K, Shimizu K, Nakashima N, Yokoyama A (1993) A new modification method of exploded lignin for the preparation of a carbon fibre precursor. J Appl Polym Sci 48(8):1485–1491

Thostensona ET, Renb Z, Choua TW (2001) Advances in the science and technology of carbon nanotubes and their composites: a review. Compos Sci Technol 61:1899–1912

Tran PA, Zhang L, Webster TJ (2009) Carbon nanofibres and carbonnanotubes in regenerative medicine. Adv Drug Deliv Rev. 61(12):1097–114. doi:10.1016/j.addr.2009.07.010

van der Woude HV, de Groot S, Janssen TWJ (2006) Manual wheelchairs: research and innovation in rehabilitation, sports, daily life and health. Med Eng Phys 2006(28):905–915

Vignoles GL, Aspa Y, Quintard M (2010) Modelling of carbon–carbon composite ablation in rocket nozzles. Compos Sci Technol 2010(70):1303–1311

Warren CD, Paulauskas FL, Baker FS, Eberle CC, Naskar A (2009) Development of commodity grade, lower cost carbon fibre—commercial applications. SAMPE J 45(2):24–36

Watt W (1985) In: Kelly A, Rabotnov YN (eds) Handbook of composites, vol I. Elsevier Science, Holland, pp 327–387

Wood K (2010) Carbon fibre reclamation: going commercial. High Perform Compos 2010(3):1–2

Zhang X, Shen Z (2002) Carbon fibre paper for fuel cell electrode. Fuel 81:2199–2201. https://www1.eere.energy.gov/vehiclesandfuels/pdfs/lm…/7_low-cost_carbon_fibre

# Chapter 2
# Lignin

**Abstract** Information on lignin is presented in this chapter.

**Keywords** Lignin · Hardwood · Softwood · Phenylpropane · Guaiacyl · Syringyl · Lignin isolation · Lignin sources

The term lignin is derived from the Latin word Lignum for wood and was first mentioned by Augustin Pyramus de Candolle, a Swiss botanist in 1819 (de Candolle and Sprengel 1821). Lignin is a complex organic polymer that forms important structural materials in the support tissues of vascular plants. It is highly polymerized and is particularly common in woody plants. The cellulose walls of the wood become impregnated with lignin. This process is called lignifications. It greatly increases the strength and hardness of the cell and imparts necessary rigidity to the tree. This is essential to woody plants in order that they stand erect (Rouhi and Washington 2001). It is one of the most abundant organic polymers on Earth after cellulose. Natural lignin is a three-dimensional polymer that occurs in many plants at levels from 15 to 32 wt% (Table 2.1). It has a complex structure containing both aromatic and aliphatic entities (Nordström 2012). Knowledge of lignin has evolved over one hundred years and the importance of lignin has been widely recognized since the early 1900s (Glasser et al. 2000). Our understanding of lignin is limited due to its complex structure. In the recent years, through the application of modern methods of chemical analysis, the lignin field has developed dramatically. This has lead to the knowledge of the structure of lignin and also to the applications of lignin. Lignin has been described as a random, three-dimensional network polymer comprised of variously linked phenylpropane units. Lignin comprises 15–25% of the dry weight of woody plants. This macromolecule plays a vital role in providing mechanical support to bind plant fibres together. Lignin also decreases the permeation of water through the cell walls of the xylem, thereby playing an intricate role in the transport of water and nutrients. Finally, lignin plays an important function in a plant's natural defense against degradation by impeding penetration of destructive enzymes through the cell wall (Sarkanen and Ludwig 1971; Sjöström 1993). Although lignin is necessary to trees, it is undesirable in

© The Author(s) 2017                                                                                    11
P. Bajpai, *Carbon Fibre from Lignin*, SpringerBriefs in Materials,
DOI 10.1007/978-981-10-4229-4_2

**Table 2.1** Composition of biomass

| Biomass | Cellulose | Hemicellulose | Lignin | Extractives |
|---|---|---|---|---|
| Softwood | 45 ± 2% | 30 ± 5% | 20 ± 4% | 5 ± 3% |
| Hardwood | 42 ± 2% | 27 ± 2% | 28 ± 3% | 3 ± 2% |
| Switchgrass | 37 ± 2% | 29 ± 2% | 19 ± 2% | 15 ± 2% |

Based on Sjöström (1993) and Mani et al. (2006)

most chemical papermaking fibres and is removed by pulping and bleaching processes.

Plant lignins can be broadly divided into three classes: softwood (gymnosperm), hardwood (angiosperm) and grass or annual plant (graminaceous) lignin (Pearl 1967). Three different phenylpropane units, or monolignols, are responsible for lignin biosynthesis (Freudenberg and Neish 1968). Guaiacyl lignin is composed principally of coniferyl alcohol units, while guaiacyl-syringyl lignin contains monomeric units from coniferyl and sinapyl alcohol. In general, guaiacyl lignin is found in softwoods while guaiacyl-syringyl lignin is present in hardwoods. Graminaceous lignin is composed mainly of p-coumaryl alcohol units.

Lignins are particularly important in the formation of cell walls, particularly in wood and bark, because they impart rigidity and do not rot easily. Lignin fills the spaces in the cell wall between cellulose, hemicellulose and the pectin components, particularly in xylem tracheids, sclereid cells and vessel elements. It is linked to hemicellulose by covalent bonds and so it crosslinks different plant polysaccharides, imparting mechanical strength to the cell wall and by extension the plant as a whole. It is especially abundant in compression wood but is rare in tension wood which are types of reaction wood. Lignin plays a very important role in conducting water in plant stems and also plays a significant role in the carbon cycle and sequesters atmospheric carbon into the living tissues of woody perennial vegetation. Lignin slowly decomposes components of dead vegetation and contributes a major fraction of the material that becomes humus as it decomposes. Lignin is a cross-linked racemic macromolecule with molecular masses in excess of 10,000 u. It is relatively hydrophobic and aromatic in nature. Lignin is polymers of phenylpropane units. Several aspects in the chemistry of lignin is still not clear.

Wood consists of 20–35% of lignin, and lignin is an important factor making wood an extraordinary material (Henriksson 2007). It is a hydrophobic material, making the cell wall impermeable to water, and thus ensuring an efficient water- and nutrition transport in the cells. The compact lignin-rich structure of wood also protects the polysaccharides from harmful microorganisms (Henriksson 2007). Wood lignin is a branched three-dimensional macromolecule mainly built up from two monolignolic units, coniferyl alcohol and sinapyl alcohol (Brodin 2009). These units give two different types of phenyl propanes in the macromolecule; guaiacyl and syringyl, respectively, which are connected with ether and carbon–carbon bonds (Sjöström 1993). Softwood lignin only contains guaiacyl units in which one of the ortho positions next to the phenol is free. The distribution of linkages in the

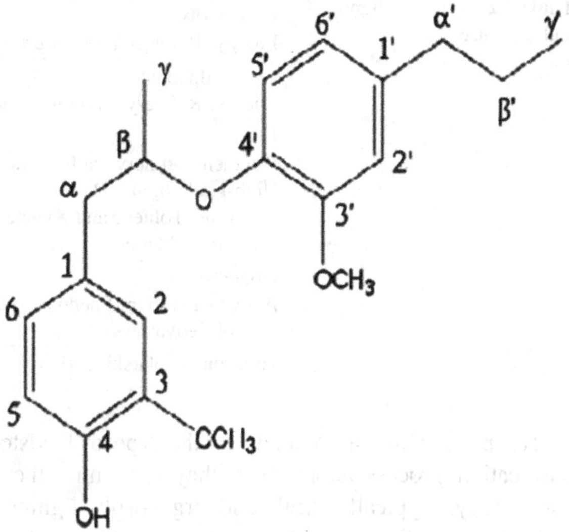

**Fig. 2.1** Common bond in lignin. The most important bond in lignin is the ß-o-4. Based on Henriksson (2007)

lignin structure appears to be random and no contrary evidence has been reported (Fig. 2.1) (Henriksson 2007; Brodin 2009).

Hardwood lignin contains syringyl units in addition to guaiacyl, which contributes to retaining a more linear structure during technical delignification and isolation processes thereof, whereas lignins originating from softwoods are more easily branched and/or cross-linked (Kubo et al. 1997; Brodin 2009). The lignin yield and the final chemical structure are also affected by the isolation methods (Sjöström 1993) and therefore affect the thermal behaviour of lignins. About two-thirds of the lignin bonds are of ether type and about one-third is carbon–carbon linkages. The most common linkage is the ß-o-4′ linkage, which represents 50–60% of the total linkages depending on the tree type (Sjöström 1993; Henriksson 2007). The β-O-4′ linkage is susceptible for pulping and bleaching and biodegrading reactions whereas the covalent carbon–carbon bonds are more stable. The functional groups in lignin are mainly methoxyl groups, phenolic hydroxyl groups and some terminal aldehyde groups. The free phenols represent only 10–13% of the total amount of aromatic rings (Henriksson 2007).

In order to use lignin as raw material, lignin has to be isolated from lignocellulose. Lignin can be derived from various sources such as cereal straws, bamboo, bagasse and wood. In terms of weight, the lignin content in wood is the highest, roughly 20–35% while other sources only contain around 3–25% (Smolarski 2012).

Around 50 million tonnes of lignin are produced annually from the pulping process; however, only approximately 1 million tonnes are isolated and sold each year for industrial applications. In general, these markets are low volume niche applications (Lucintel Market Report 2011; Luo 2010). They can be divided in three main groups, Biofuel, macromolecules and aromatics (NNFCC 2009; Chapple et al. 2007; Smolarski 2012).

**Table 2.2** Types of lignin
and their use

| |
|---|
| Low purity |
| Energy, Refinery (carbon cracker) |
| Lignosulphonates |
| Energy, Refinery (carbon cracker) |
| Kraft |
| Bitumen, Refinery (carbon cracker) , Cement additives, Biofuel, High-grade lignin, BTX (Benzene, Toluene and Xylene), Phenolic resins, Carbon fibres, Vanillin, Phenol |
| Organosolv |
| Activated carbon, Phenolic resins, Carbon fibres, Vanillin, Phenol derivatives |
| Based on Smolarski (2012) |

The price of lignin depends on the type of feedstock and the degradation and purification process adopted, as they determine the lignin structure, purity and consistency. Typically, kraft and organosolv lignins are suitable candidates for high-value applications whereas lingosulphonate lignin most likely is used for lower value products, Table 2.2. Additionally, kraft lignin covers several applications, including the high-value ones, and is considered mid-range in terms of price. It is also readily available in sufficient quantity from pulp and paper manufacturers to start meeting the industrial demand and is considered a good intermediate between lignosulphonates and organosolv lignin.

# References

Brodin I, Sjöholm E, Gellerstedt G (2009) Kraft lignin as feedstock for chemical products: the effects of membrane filtration. Holzforschung 63:290–297

Chapple C, Ladisch M, Meilan R (2007) Loosening lignin's grip on biofuel production. Nat Biotechnol 25(7):746–748. doi:10.1038/nbt0707-746

de Candolle AP, Sprengel KPJ (1821) Elements of the philosophy of plants: containing the principles of scientific botany … with a history of the science, and practical illustrations (trans: AP de candolle, KPJ Sprengel, from the German). W. Blackwood, Edinburgh

Freudenberg K, Neish AC (1968) Constitution and biosynthesis of lignin. In: Springer GF, Kleinzeller A (ed). Springer, New York, 129 pp

Glasser WG, Northey RA, Schultz TP (2000) Lignin: historical, biological, and materials perspective. Am Chem Soc. Washington, DC

Henriksson G (2007) In Ljungberg textbook. Pulp and paper chemistry and technology. In: Ek M, Gellerstedt G, Henriksson G (eds) Fibre and polymer technology, KTH, Stockholm, 2007, Book 1, pp 125–148

Kubo S, Ishikawa M, Uraki Y, Sano Y (1997) Preparation of lignin fibres from softwood acetic acid lignin relationship between fusibility and the chemical structure of lignin. Mokuzai Gakkaishi 43:655–662

Lucintel Market Report (2011) Growth opportunities in the global carbon fibre market: 2011–2016. Lucintel, Las Colinas

Luo J (2010) Lignin-based carbon fibre. Thesis Master of Science (in Chemical Engineering), The University of Maine, May, 2010

Mani S, Tabil LG, Sokhansanj S (2006) Effects of compressive force, particle size and moisture content on mechanical properties of biomass pellets from grasses. Biomass Bioenergy 30:648–654

NNFCC (2009) Marketing study for biomass treatment technology (No. NNFCC Project 10/003). Heslington, UK. http://www.nebr.co.uk/_cmslibrary/files/marketingstudynnfccappendix2.pdf

Nordström Y (2012) Development of softwood kraft lignin based carbon fibres. Licentiate Thesis, Division of Material Science Department of Engineering Sciences and Mathematics, Luleå University of Technology

Pearl IW (1967) The chemistry of lignin. Marcel Dekker, Inc., New York, 339 pp

Rouhi AM, Washington C (2001) Only facts will end lignin war. Sci Technol 79(14):52–56

Sarkanen KV, Ludwig CH (1971). Lignin: occurrence, formation, structure and reactions. In: Sarkanen KV, Ludwig CH (ed). Wiley-Interscience, New York, 916 pp

Sjöström E (1993) Wood chemistry: fundamentals and application. Academic Press, Orlando, p 293

Smolarski N (2012) High-value opportunities for lignin: unlocking its potential. Frost & Sullivan. www.greenmaterials.fr/wp-content/uploads/2013/01/Highvalue-Opportunities-for-Lignin-Unlocking-its-Potential-Market-Insights.pdf

# Chapter 3
# Carbon Fibre

**Abstract** Properties of carbon fibre and its processing are presented in this chapter.

**Keywords** Carbon fibre · Mechanical properties · Specific modulus · Carbonization · Graphitization · Price of carbon fibre

Carbon fibre is defined as a fibre containing at least 90 wt% carbon while the fibre containing at least 99 wt% carbon is usually called a graphite fibre. Carbon fibre possesses a unique combination of properties and is used in the manufacture of advanced composite materials (Baker and Rials 2013). The material comes in a variety of "raw" building-blocks, including yarns, uni-directional, weaves, braids, and several others. Carbon fibre is lightweight, has high strength, flexibility and fatigue resistance. The properties of carbon fibre (lightweight, high strength, flexibility and fatigue resistance) vary widely depending on the structural orientation of the fibre axis (Luo 2010).

In Table 3.1, mechanical properties and specific gravity of different materials are presented for comparison (Peebles 1995; Bunsell and Renard 2005; Hull and Clyne 1996; Ventura and Martelli 2009). When compared with metals like steel, carbon fibres possess an order of magnitude higher specific modulus and strength. When compared on an absolute basis with strong polymer, i.e. Kevlar fibres, they display much higher modulus and strength. Overall, carbon fibres possess an excellent combination of modulus, strength, and conductivities as compared with other materials.

It is manufactured by thermally treating a precursor fibre in a process termed "carbonization". Formation of the precursor fibre can be done either by melt spinning or wet spinning. In the melt spinning process, the precursor raw material is simply melted and extruded through an orifice to form the fibre. In the wet spinning process, the precursor material is dissolved in a suitable solvent and then extruded through the orifice to form the fibre. Before carbonization, the precursor fibre is subjected to a process termed "thermo-stabilization". The thermo-stabilization process causes cross-linking of the polymer on the fibre surfaces; thus preventing shrinking, melting, and fusing.

P. Bajpai, *Carbon Fibre from Lignin*, SpringerBriefs in Materials,
DOI 10.1007/978-981-10-4229-4_3

**Table 3.1** Properties of carbon fibres and its comparison with other materials

| Material | Specific gravity | Modulus (GPa) | Specific modulus (GPa) | Electrical resistivity ($\mu\Omega m$) | Thermal conductivity (W/m-K) | Strength (GPa) | Specific strength (GPa) |
|---|---|---|---|---|---|---|---|
| Carbon fibres (high strength) | 1.82 | 294 | 164 | N/A | N/A | 7.1 | 3.9 |
| Kevlar fibres | 1.45 | 135 | 93 | N/A | 0.04 | 2–3 | ~1.5–2 |
| Steel | 7.9 | 200 | 25 | 0.72 | 50 | 1–1.5 | 0.1–0.2 |
| Glass (bulk and fibre) | 2.5 | 72 | 28 | $10_{18}$–$10_{19}$ | 13 | 2-3 | ~1–1.5 |
| Aluminium | 2.7 | 76 | 28 | 0.003 | 205 | 0.5 | 0.2 |

Zhang (2016), Peebles (1995), Bunsell and Renard (2005), Hull and Clyne (1996),Ventura and Martelli (2009)

Based on its mechanical properties, carbon fibre is classified into two groups: general purpose or high performance. The precursor materials used to produce carbon fibre are extremely important in determining the final properties and its classification. Pitch, derived from petroleum or coal, and polyacrylonitrile (PAN) are the most important types of precursor materials used to produce carbon fibre commercially. Almost 80% of commercial carbon fibre is predicated on using PAN as the starting raw material because of its superior properties compared to those of pitch-based carbon fibre (Luo 2010). Carbon fibre produced from PAN is expensive, and thus has limited application to high-performance structural materials. There is a need for a low-cost precursor material that can produce carbon fibre with properties superior to those of pitch and approaching those of PAN. The carbon content in commercial carbon fibre must be at least 92% carbon by weight.

Carbon fibre comes in a variety of types. They can be short or continuous; their structure can be crystalline, amorphous, or partly crystalline (Chung 1994). Carbon fibre has a high modulus of elasticity that results from the fact that the carbon layers tend to be parallel to the fibre axis. Fibre "texture" is a term applied to this preferred orientation the crystal structure. The modulus of elasticity of carbon fibre is higher parallel to the fibre axis than perpendicular to the axis. The stronger the "fibre texture", the greater the degree of alignment of the carbon layer parallel to the fibre axis. Carbon fibre with high fibre texture has high strength and high tensile energy absorption (TEA). Carbon fibre generally has excellent tensile strength, low densities, high thermal and chemical stabilities in the absence of oxidizing agents, good thermal and electrical conductivities and excellent creep resistance. Carbon fibre generally comes in the form of woven textile, pre-preg, continuous fibres, rovings and chopped fibres. To further process the composite parts into final product, the composite parts can be produced through filament winding, tape winding, pultrusion, compression moulding, vacuum bagging, liquid moulding or injection moulding (Huang 2009).

Carbon fibre is grouped into a wide range of categories based on its modulus of elasticity (Price 2011). Carbon fibres classified as "low modulus" have a tensile modulus below 34.8 million psi (240 million kPa). Other classifications, in ascending order of tensile modulus, include "standard modulus", "intermediate modulus", "high modulus" and "ultra-high modulus". Ultra-high modulus carbon fibres have a tensile modulus of 72.5–145.0 million psi (500 million–1.0 billion kPa). As a comparison, steel has a tensile modulus of about 29 million psi (200 million kPa). Thus, the strongest carbon fibres are ten times stronger than steel and eight times that of aluminium, not to mention much lighter than both materials, 5 and 1.5 times, respectively. Additionally, their fatigue properties are superior to all known metallic structures, and they are one of the most corrosion-resistant materials available, when coupled with the proper resins.

Price also varies dramatically among different categories. High and ultra-high stiffness fibre is made for the aerospace industry at around $2000 per kg. They are expensive and used in specialized applications such as airfoils. Standard modulus fibre can be as low as $22 per kg for the civil infrastructure industry. Table 3.2 shows the classification of carbon fibre. Table 3.3 shows properties of carbon fibre and Table 3.4 shows some disadvantages of carbon fibre.

| Table 3.2 Classification of carbon fibre | |
|---|---|
| | Low Modulus<br>Modulus of Elasticity—40–200 GPa<br>Less than $20/kg<br>Non-structure usage |
| | Standard modulus<br>Modulus of elasticity—200–275 GPa<br>$20/kg–$55/kg<br>Automotive, sporting goods, wind turbine, pressure tanks |
| | Intermediate modulus<br>Modulus of elasticity—275–345 GPa<br>$55/kg–$65/kg<br>Pressure tanks, wind turbine |
| | High modulus<br>Modulus of elasticity—345–600 GPa<br>$65/kg–$90/kg<br>Aviation, military |
| | Ultra-high modulus<br>Modulus of elasticity—600–965 GPa<br>Up to $2000/kg<br>Aerospace, Military |
| | Based on Chen (2014), Eberle (2012), Price (2011) |

**Table 3.3** Properties of carbon fibre

| Low density |
| --- |
| High tensile modulus and strength |
| Low thermal expansion coefficient |
| Thermal stability in the absence of oxygen over 3000 °C |
| Excellent creep resistance |
| Chemical stability, particularly in strong acid |
| Biocompatibility |
| High thermal conductivity |
| Low electrical resistivity |
| Chung (1994) |

**Table 3.4** Disadvantages of carbon fibre

| High cost (currently) |
| --- |
| Anisotropy (in the axial versus transverse direction) |
| Compressive strength is low compared to tensile strength |
| Oxidation of carbon fibre is catalysed by an alkaline environment |
| Chung (1994) |

Carbon fibre is used as a reinforcing material in composite products. Industrial use of carbon fibre began in the 1950s in the production of aircraft and aerospace materials because of its unique properties especially the low density, high modulus and fatigue resistance (Kadla et al. 2002). Because of its excellent properties, today carbon fibre is widely used in diverse applications such as tennis rackets, bicycles, fishing poles, boats and high-performance jet aircraft to name a few. In the modern automobile industry, a priority of automobile manufacturers is to develop energy-saving vehicles. Carbon fibre composites could replace traditional steel components and greatly decrease the weight of vehicles resulting in substantial increase in fuel efficiency. With economic development, the superior performance of carbon fibre composite can be applied to many more products. However, to make this a reality the price of carbon fibres must be reduced substantially by reducing the cost of the raw materials and simplification of the manufacturing process.

High-quality carbon fibre is currently manufactured from PAN and is derived from propylene and ammonia, which come from crude oil, nitrogen and natural gas. PAN has been extensively researched and is in commercial production (Kadla et al. 2002). Lower quality material is produced from petroleum pitch and coal tar.

Like all man-made fibres manufactured from polymeric precursors, carbon fibre is manufactured in several common steps starting with the raw polymer. The processing involves

(1) spinning or extrusion of the polymer,
(2) stabilization or conversion of the spun fibre into a stable chemical form,

**Fig. 3.1** Process for
manufacture of carbon fibre
from PAN

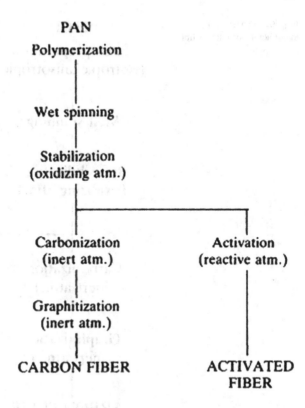

(3) carbonization or conversion of the chemical structure into carbon,
(4) graphitization or conversion of the carbon into graphite which has a distinct
    structure. Starting with pitch as the precursor, a process very similar to that used
    for PAN is employed except that melt spinning is used in place of wet spinning.
    The processes for using PAN or pitch as precursors to produce carbon fibre are
    shown in Figs. 3.1 and 3.2 (Chung 1994). Norberg et al. (2013) estimates those
    lignin fibre costs at $1.1/kg, whereas Baker et al. (2012) predicts a price for one
    kilogram as low as $0.85 but both estimate the same cost of $6.2/kg for the
    finished carbon fibre. Thus, a focus on carbon fibre from low-cost and sus-
    tainable materials could result in significantly lower final costs (Baker 2009;
    Attwenger 2014).

**Fig. 3.2** Process for
manufacture of carbon fibre
from Pitch

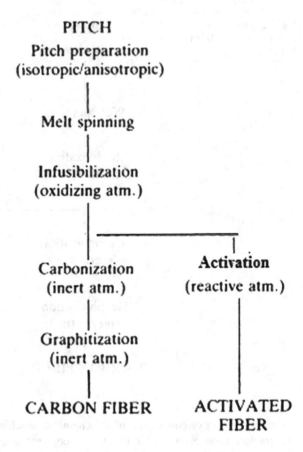

**PITCH**

**Pitch preparation
(isotropic/anisotropic)**

**Melt spinning**

**Infusibilization
(oxidizing atm.)**

**Carbonization
(inert atm.)**        **Activation
(reactive atm.)**

**Graphitization
(inert atm.)**

**CARBON FIBER**        **ACTIVATED
FIBER**

# References

Attwenger A (2014) Value-added lignin-based carbon fibre from organosolv fractionation of poplar and switchgrass. A thesis presented for the Master of Science Degree, The University of Tennessee, Knoxville

Baker FS (2009) Low cost carbon fibre from renewable resources. Oak Ridge, TN. Retrieved from http://www1.eere.energy.gov/vehiclesandfuels/pdfs/merit_review_2009/light

Baker DA, Rials TG (2013) Recent advances in low-cost carbon fibre manufacture from lignin. J Appl Polym Sci 130:713–728

Baker DA, Gallego NC, Baker FS (2012). On the characterization and spinning of an organic-purified lignin toward manufacture of low-cost carbon fibre. J Appl Polym Sci 227–234

Bunsell AR, Renard J (2005) Fundamentals of fibre reinforced composite materials. IOP Publishing Ltd., London, pp 3–5

Chen MCW (2014) Commercial viability analysis of lignin based carbon fibre. Master thesis, Simon Fraser University

Chung DL (1994) Carbon fibre composites. Butterworth-Heinemann, New York, p 1994

Eberle C (2012) Carbon fibre from lignin. Oak Ridge, TN. Retrieved from http://www. cfcomposites.org/PDF/Breakout_Cliff.pdf

Huang X (2009) Fabrication and properties of carbon fibres. Materials 2(4):2369–2403. doi:10.3390/ma2042369

Hull D, Clyne T (1996) An introduction to composite materials. Cambridge University Press, Cambridge

Kadla JF, Kubo S, Gilbert RD, Venditti RA (2002) Lignin-based carbon fibres. In Chemical modification, properties, and usage of lignin. In Hu TQ, (ed) Kluwer Academic/Plenum, New York, pp 121–137

Luo J (2010) Lignin-based carbon fibre. A thesis, Degree of Master of Science in Chemical Engineering, The University of Maine, Maine

Norberg I, Nordström Y, Drougge R, Gellerstedt G, Sjöholm E (2013) A new method for stabilizing softwood kraft lignin fibres for carbon fibre production. J Appl Polym Sci 128:3824–3830

Peebles L (1995) Carbon fibres, formation, structure and properties, Carbon fibres, formation, structure and properties. CRC Press, Boca Raton, pp 7–24

Price RE (2011) Carbon Fibre used in Fibre Reinforced Plastic (FRP). Prince engineering. Retrieved June 11, 2014, from http://www.build-on-prince.com/carbon-fibre.html

Ventura G, Martelli V (2009) Thermal conductivity of Kevlar 49 between 7 and 290K. Cryogenics 49:735–737

Zhang M (2016) Carbon fibres derived from dry-spinning of modified lignin precursors Ph.D. thesis, Chemical engineering, Clemson University, Clemson

# Chapter 4
# Carbon Fibre Market

**Abstract** The global carbon fibre market and demand and the major growth drivers are presented in this chapter.

**Keywords** Carbon fibre market · Carbon fibre demand · Carbon fibre capacity · Industrial applications · Carbon fibre reinforced plastic · PAN-based carbon fibre · Pitch-based carbon fibre

Carbon fibre is becoming the material of choice for manufacturers around the world. The future of the global carbon fibre market appears to be attractive with opportunities in several areas—industrial applications, sporting goods and aerospace (Fitzer 1990; Chung 1994; Watt 1985; Donnet and Bansal 1990; Minus and Kumar 2005; Minus and Kumar 2007; Fitzer et al. 1989; Hajduk 2005; Huang 2009).

According to Transparency Market Research report (www.transparencymarket research.com 'Chemical & Materials' Carbon Fibre Market), the global carbon fibre market is expanding progressively at a CAGR of 9.9% from 2014 to 2020. In 2013, the global value of the carbon fibre market was US$1.94 billion and is expected to reach a value of US$3.73 billion by 2020.

Other market reports predict that the global carbon fibre market will grow at a CAGR of 11.0% by value from 2016 to 2021 (www.transparencymarketresearch.com 'Chemical & Materials' Carbon Fibre Market https://globenewswire.com/.../2016/ .../Growth-Opportunities-in-Global-Carbon-Fibre-...; www.researchandmarkets. com/reports/3743269/growth-opportunities-in-global-carbon-fibrehttp://www.mark etresearch.com/Lucintel-v2747/Growth-Opportunities-Global-Carbon-Fibre-1009 3344/; https://www.highbeam.com/doc/1G1-454852745.html) (Table 4.1).

According to Kraus et al. (2015), total turnover achieved worldwide with carbon fibre makes up around 1.98 billion US$ for 2014. In 2013, it was 1.77 billion US$ (2013), showing a growth rate of 11.9%. Overall, the demand for carbon fibre has shown steady growth since the general economic recession of 2009.

The main growth drivers are increasing demand of lightweight material in end-use industries. The other main drivers are increasing use of carbon fibre in the

© The Author(s) 2017

P. Bajpai, *Carbon Fibre from Lignin*, SpringerBriefs in Materials,
DOI 10.1007/978-981-10-4229-4_4

**Table 4.1** Global carbon fibre demand in thousand tones

| | |
|---|---|
| 2008 | 31.5 |
| 2009 | 26.5 |
| 2010 | 33.0 |
| 2011 | 38.5 |
| 2012 | 43.5 |
| 2013 | 46.5 |
| 2014 | 51.0 |
| 2015 | 58.0 |
| 2016 | 65.0 |
| 2019 | 92.0 |
| 2022 | 116.0 |

Based on Kraus et al. (2015)

automotive, wind turbine blades, commercial aircraft (Boeing 787 and Airbus 380) and several industrial applications. In the next 5 years, the demand for industrial application is expected to experience the highest growth, supported by increasing demand for lightweight materials in automotive industry and the growing wind energy. Within global carbon fibre market, it is expected that industrial applications will remain as the largest market by volume consumption. Growing demand of lightweight materials in the automotive industry and increased wind turbine blade length is anticipated to stimulate growth for this segment over the next five years.

The carbon fibre market can be divided into two types based on the precursors type:

- PAN (polyacrylonitrile) -based
- Pitch-based.

Use of PAN-based carbon fibre is higher by volume and value because of its lower cost than pitch-based. Because of the increasing demand of lightweight materials in the aerospace and automotive industries and growth in end-use industries, North America is expected to remain the largest market. Rest of World, including Asia Pacific is expected to witness the highest growth because of the anticipated growth in the end-use industries and increasing focus on high performance composite materials. Toho Tenax, Toray, SGL, Cytec, Hexcel, Mitsubishi Rayon are among the major suppliers of the carbon fibre.

According to Lucintel, the demand for industrial application is expected to experience the highest growth in the next five years, supported by growing wind energy industry and increasing demand for lightweight materials in automotive industry. According to Acmite, global revenues for the carbon fibre market in 2013 totalled US$1.77 billion. In 2012, this figure was US$1.63 billion.

Carbon fibres are known for its high strength to weight ratio. So they are used for manufacturing of aircraft components such as wings, elevators, floor beams, engine

**Table 4.2** Carbon fibre capacity by manufacturer in 1000 tonnes (2014)

| | |
|---|---|
| Toray and Zoltek | 21 |
| Sgl | 9.0 |
| Toho | 11.5 |
| MRC | 11.1 |
| Form Plastic Cor. | 8.8 |
| Hexel | 7.2 |
| Cytec | 4.0 |
| Zhongfu-Shenying | 4.0 |
| Hengshen fibre material | 3.0 |
| Aksa | 1.8 |
| Others | 9.2 |

Based on Kraus et al. (2015)

nacelles and vertical stabilizers among others (www.credenceresearch.com/report/carbon-fibre-market). Hence, it is expected that carbon fibres industry will experience high demand from various end user industries globally during the next 5 years.

In terms of value, the global carbon fibre market size is projected to reach USD 3.51 Billion by 2020, at a CAGR of 9.1% between 2015 and 2020, whereas the global carbon fibre reinforced plastic (CFRP) market is predicted to reach USD 35.75 Billion by 2020, at a CAGR of 9.9% between 2015 and 2020 www.marketsandmarkets.com/PressReleases/carbon-fibre-composites.asp.

Toray, Zoltek, Toha, MRC, SGL Form Plast Cor. Hexcel, Hyosung Aksa Diverse are the leading carbon fibre producers. These ten leading fibre manufacturers make up almost 88% of the global carbon fibre capacity and are still the predominant forces. When the carbon fibre demand from 2014 is taken into consideration, then over capacity is still lying at around 42% (Table 4.2).

Toray has taken over Zoltek. The acquisition was acquired in 2014. These two firms together had a production capacity of 44,500 tonnes of carbon fibre yearly in 2014. Toray increased its capacity by 6000 tonnes to 27,100 tonnes in 2014 (Toray 2012, 2015). With an estimated total global capacity of 125,000 tonnes of carbon fibre on the basis of Polyacrylnitrile (PAN) and Pitch, this includes around one-third of the global carbon fibre market. Other fibre manufacturers have also increased their carbon fibre production capacities, or have a plan to increase. Two subsidiaries of Mitsubishi Chemical Holdings Corporation-Mitsubishi Rayon (MRC) and Mitsubishi Plastics (MPI) will operate together in the carbon fibre business in the future. MRC produces PAN-based Carbon fibres and Pitch-based fibres (Mitsubishi Plastics 2015). The carbon fibre capacity of MRC has been estimated at 11,100 tonnes and a further expansion of the production capacity in Sacramento in USA has been announced (Mitsubishi Rayon 2014). SGL and BMW have installed two further production lines in Moses Lake having a capacity of 3000 tonnes (BMW Group 2014).

# References

BMW Group, BMW Group PresseClub Deutschland, 09 05 2014. (Online). Available: https://www.press.bmwgroup.com/deutschland/

Chung DL (1994) In carbon fibre composites. Butterworth-Heinemann, Boston, pp 3–11

Donnet JB, Bansal RC (1990) Carbon fibres, 2nd edn. Marcel Dekker, New York, pp. 1–145

Fitzer E, Edie DD, Johnson DJ (1989) In carbon fibres filaments and composites, 1st edn. In: Figueiredo JL, Bernardo CA, Baker RTK, Huttinger KJ (eds) Springer, New York, pp 3–41, 43–72, 119–146

Fitzer E (1990) In: Figueiredo JL, Bernardo CA, Baker RTK, Huttinger KJ (eds) Carbon fibres filaments and composites. Kluwer Academic, Dordrecht, pp 3–4

Hajduk F (2005). Carbon fibres overview. Global outlook for carbon fibres 2005, Intertech conferences. San Diego, CA, 11–13 Oct 2005

Huang H (2009) Fabrication and properties of carbon fibres. Materials 2:2369–2403

Kraus T, Kühnel M, Elmar Witten E (2015) Composites market report 2015

Market developments, trends, outlook and challenges (2014) www.eucia.eu/userfiles/files/20141008_market_report_grpcrp.pdf

Minus ML, Kumar S (2005) The processing, properties, and structure of carbon fibers. JOM 57(2):52–58

Minus ML, Kumar S (2007) Carbon fibre. Kirk-Othmer Encycl Chem Technol 26:729–749

Mitsubishi Plastics, Inc., Mitsubishi Rayon Co., Ltd. (2015) News release: enhancement of the carbon fibre business, 07 01 2015. (Online). Available: http://www.mpi.co.jp/english/news/201501070751.html

Mitsubishi Rayon, Mitsubishi Rayon Pressroom, 30 06 2014. (Online). Available: https://www.mrc.co.jp/english/pressroom/detail/pdf/20140630192937.pdf

Toray Global, Production Capacity Toray Group, 04 2015. (Online). Available: http://www.toray.com/ir/management/man_010.html

Toray Global, Press Releases Toray Group, 09 03 2012. (Online). Available: http://www.toray.com/news/crb/nr120309.html

Watt W (1985) In: Kelly A, Rabotnov YN (eds) Handbook of composites—volume I, Elsevier Science, Holland, pp 327–387

# Chapter 5
# Lignin as a Raw Material for Carbon Fibre

**Abstract** Lignin as a raw material for carbon fibre is presented in this chapter. In contrast to PAN and petroleum pitch, lignin is a renewable material and has a carbon content of more than 60%. Furthermore, it is available in large quantities; possible to isolate and eventually modify, and still obtainable as a relatively cost-competitive raw material, as compared with PAN. Cost estimations for lignin as precursor shows potential of a remarkably cost reduction as compared to PAN based precursor.

**Keywords** Lignin · Renewable material · Carbon fibre · Cost-competitive · Low cost carbon fibre

Of the carbon fibres produced today, over 90% originates from the oil-based synthetic polymer polyacrylonitrile, which is expensive to produce (Baker and Rials 2013); small amounts are produced from pitches also. But, still carbon fibre remains a specialty product due to the high cost of these petroleum-based precursors and their processing costs. Therefore, as such has been limited for use in sporting goods, aerospace, high-end automotive, and specialist industrial applications (Huang 2009; Baker et al. 2012). Research towards the manufacture of low-cost carbon fibre have been limited to a small number of organizations due to limited availability of expertise, the magnitude of the effort needed, a limited availability of expertise, equipment provision and the cost. These efforts have typically involved the reduction of cost by reducing the cost of processing, use of lower cost materials or using a combination of the two.

U.S. Department of Energy (DOE) studies have identified various materials that could be used for the lightweighting of vehicles. One of the most promising material, PAN-based carbon fibre reinforced composites, could offer as much as 60% part weight reduction, but the cost is ten times higher (Warren 2009). The DOE has made an investment well over US$100 M over the last 10 years for examining possible routes towards the provision of suitable low-cost carbon fibre, representing possibly the largest single investment towards that goal.

© The Author(s) 2017                                                                                  29
P. Bajpai, *Carbon Fibre from Lignin*, SpringerBriefs in Materials,
DOI 10.1007/978-981-10-4229-4_5

To reduce carbon fibre manufacturing costs DOE emphasis has been towards reducing processing costs using the following measures (Warren et al. 2008, 2009a, b; White et al. 2006; Paulauskas et al. 2006, 2010; Paulauskas 2010; Horikiri et al. 1978):

- Advanced oxidation techniques
- Potential use of textile grade pan precursors
- Synthesis of new melt-spinnable pan precursors
- Conversion of polyolefins to carbon fibre.

Another DOE program (Baker 2010a, b; Baker et al. 2009) involves the manu-facture of carbon fibre from lignin. This is directed towards cost reduction through the very low cost of the renewable precursor. In contrast to PAN and petroleum pitch, lignin is a renewable material. Lignin has a carbon content of more than 60% (Gellerstedt et al. 2010). Furthermore, lignin is available in large quantities; pos-sible to isolate and eventually modify, and still obtainable as a relatively cost-competitive raw material, as compared with PAN. Cost estimations for lignin as precursor shows potential of a remarkably cost reduction as compared to PAN-based precursor (Baker 2010a, b). Lignin-based carbon fibre is the most value-added product from a wood-based biorefinery (Gosselink 2011).

The best lignin-based carbon fibre samples produced to date had the following properties (Warren and Naskar 2012):

- Average strength of 1.0 gpa (155 ksi; 109 kg-f/mm$^2$)
- Moduli of 82.7 gpa (12 Msi; 8.44 $\times$ 103 kg-f/mm$^2$)
- Extensibilities of 2.03%.

These samples were manufactured using a modified technical lignin and were projected to cost much less than any other known method of carbon fibre pro-duction (Baker et al. 2010a, b; Warren 2011). Attempts to further increase strength have been restricted by the unavailability of suitable lignins (lignins that meet stringent melt spinning requirements) to allow for multifilament melt spinning and conversion trials. However, there is a possibility this could change.

One of the major driving forces behind the development of lignin carbon fibre in recent years is government regulatory changes on fuel consumption. The U.S. government in 2012 legislated through updated Corporate Average Fuel Economy (CAFE) standards that the average fuel economy of cars and light trucks sold in the U.S. for model year 2017 will be 35.5 mpg, and will increase to 54.5 mpg for 2025 models. The most effective way to increase fuel economy is to reduce the vehicle weight (Berkowitz 2011). Presently, PAN-based carbon fibre reinforced composites could offer up to 60% weight reduction at ten times the cost. As more than 50% of the manufacturing cost is the cost of precursor, which fluctuates severely with the oil price, Department of Energy has invested well over $100 million over the last decade to examine possible routs of low-cost precursor alternatives. Lignin-based precursor is one of the most promising contenders (Baker and Rials 2013; Huang 2009; Kubo et al. 1997; Kadla et al. 2002; Kubo and Kadla 2005; Sudo and Shimizu 1992; Uraki et al. 1995; Nordström 2012).

Oak Ridge National Laboratory (ORNL) in United States is active in pursuing research on lignin-based carbon fibre. Oak Ridge Carbon Fibre Composites Consortium was established in 2011 to accelerate the development of low-cost carbon fibre reinforced composite materials. The objectives set by the consortium is to manufacture a lignin-based carbon fibre with a tensile strength of 1.72 GPa and a modulus of 172 GPa that is suitable for the automotive industry. The consortium is having more than 52 members across the whole carbon fibre value chain starting from raw materials to downstream applications.

The high carbon content in the lignin macromolecules makes it a potential candidate for carbon fibre production. However, a suitable precursor must also have fibre forming ability by some of the spinning methods and withstand the following high temperature treatment. Thermoplasticity has been observed for lignins indicating the possibility to use melt extrusion for fibre spinning (Kadla et al. 2002; Kubo and Kadla 2005) which is the preferred processing technique as it is less costly than wet spinning.

The desire to produce lignin based carbon fibres is not new. Already in the 1960s, a commercial carbon fibres was produced from lignin, called the Kayocarbon fibre. It was produced from lignosulphonates by dry spinning together with a plasticizer, polyvinyl alcohol (Otani et al. 1969). However, the production soon discontinued. The PAN process was improved which enabled an ever since increasing market share for PAN-based carbon fibres (Lucintel 2011). However, the drive for finding alternatives for PAN-based carbon fibres is intensifying, because of increasing demands combined with higher raw material price and environmental awareness (Warren 2009).

# References

Baker DA, Gallego NC, Baker FS (2012) On the characterization and spinning of an organic-purified lignin toward the manufacture of low-cost carbon fibre. J Appl Polym Sci 124:227–234

Baker DA, Rials TG (2013) Recent advances in low-cost carbon fibre manufacture from lignin. J Appl Polym Sci 130(2):713–728. doi:10.1002/app.39273

Baker FS (2010a) Presentation at 2010 DOE Hydrogen program and vehicle technologies annual merit review and peer evaluation meeting, June 7–11, 2010. Available at http://www1.eere.energy.gov/vehiclesandfuels/pdfs/merit_review_2010/lightweight_materials/lm005_baker_2010_ o.pdf

Baker FS (2010b) Low Cost Carbon Fibre from Renewable Resources, June 7–11, 2010. Available at http://www1.eere.energy.gov/vehiclesandfuels/pdfs/merit_review_2010/lightweight_materials/lm005_baker_2010_o.pdf2010. Oak Ridge National Laboratory

Baker FS, Gallego NC, Baker DA (2009) DOE FY 2009 progress report for lightweighting materials, Part 7.A., 2009. Available at http://www1.eere.energy.gov/vehiclesandfuels/pdfs/lm_09/7_low-cost_carbon_fibre.pdf

Baker FS, Baker, DA, Gallego NC (2010a) Carbon fibre from lignin. Proceeding, Carbon 2010, Clemson, SC, July 11–16

Baker FS, Baker DA, Gallego NC (2010b) Proceeding, SAMPE '10 Conference and Exhibition, Seattle, WA, May 17–20

Berkowitz J (2011) The CAFE numbers game: making sense of the new fuel economy regulations. Car and Driver. Retrieved June 23, 2014, from http://www.caranddriver.com/features/the-cafe-numbers-game-making-sense-of-the-newfuel-economy-regulations-feature

Gellerstedt G, Sjöholm E, Brodin I (2010) The Wood-based biorefinery: a source of carbon fibre? Open Agric J 3:119–124

Gosselink RJA (2011) Lignin as a renewable aromatic resource for the chemical industry. Ph.D. Thesis—Proefschrift Wageningen, 2011

Horikiri S, Iseki J, Minobe M (1978) Process for producing carbon fibre, US Patent 4070446

Huang X (2009) Fabrication and properties of carbon fibres. Materials 2:2369–2403. doi:10.3390/ma2042369

Kadla JF, Kubo S, Venditti RA, Gilbert RD, Compere AL, Griffith W (2002) Lignin based carbon fibres for composite fibre applications. Carbon 40:2913–2920

Kubo S, Ishikawa M, Uraki Y, Sano Y (1997) Preparation of lignin fibres from softwood acetic acid lignin relationship between fusibility and the chemical structure of lignin. Mokuzai Gakkaishi 43:655–662

Kubo S, Kadla JF (2005) Lignin-based carbon fibres: effect of synthetic polymer blending on fibre properties. J Polym Environ 13:97–105

Lucintel, Market Report (2011) Growth opportunities in the global carbon fibre market, 2011–2016. Lucintel, Las Colinas

Nordström Y (2012) Development of softwood kraft lignin based carbon fibres, Licentiate Thesis. Department of Engineering Sciences and Mathematics Luleå University of Technology

Otani S, Fukuoka Y, Igarashi B (1969) Methods for Producing Carbonized Lignin Fibre. US Patent 3461082

Paulauskas FL, White TL, Spruiell JE (2006) Proceedings of the International SAMPE Technical Conference, Long Beach, CA, USA, May 2006

Paulauskas FL (2010) Presentation at 2010 DOE Hydrogen program and vehicle technologies annual merit review and peer evaluation meeting, June 9, 2010. Available at http://www.hydrogen.energy.gov/pdfs/review10/st093_paulauskas_2010_p_web.pdf

Paulauskas FL, Norris R, Naskar A, Ozcan S, Yarborough K (2010) DOE Hydrogen program annual progress report, 2010, 622–627. Available at http://www.hydrogen.energy.gov/pdfs/progress10/iv_g_3_norris.pdf

Sudo K, Shimizu K (1992) A new carbon fibre from lignin. J Appl Polym Sci 44(1):127–134

Uraki Y, Kubo S, Nigo N, Sano Y, Sasaya T (1995) Preparation of carbon fibres from organosolv lignin obtained by aqueous acetic acid pulping. Holzforschung 49:343–350

Warren CD, Paulauskas FL, Baker FS, Eberle CC, Naskar A (2008) Multi-task research program to develop commodity grade, lower cost carbon fibre. In: Proceedings of the SAMPE Fall Technical Conference, Memphis, TN, USA, Sept 2008

Warren CD, Paulauskas FL, Baker FS, Eberle CC, Naskar A (2009a) Development of lower cost carbon fibre for high volume applications. SAMPE J 45(24–36):33

Warren CD, Paulauska FL, Eberle CC, Naskar AK, Ozcan S (2009b) Proceedings of the 17th Annual International Conference on Composites/Nano Engineering, Hawaii, USA, July 2009

Warren DC (2009) Low cost carbon fibre research in the LM materials program overview. Oak Ridge, TN. Retrieved from https://www1.eere.energy.gov/vehiclesandfuels/pdfs/merit_review_2009/lightweight_materials/lm_02_warren.pdf

White SM, Spruiell JE, Paulauskas FL (2006) Fundamental studies of stabilization of polyacrylonitrile precursor, part 1. Effects of thermal and environmental treatments Proceedings of the International SAMPE Technical Conference, Long Beach, CA, USA, May 2006

Warren CD, Naskar AK (2012) Presentation at 2012 DOE Hydrogen and fuel cells program and vehicle technologies program annual merit review and peer evaluation meeting, May 16, 2012. Available at http://www1.eere.energy.gov/vehiclesandfuels/pdfs/merit_review_2012/lightweight_materials/lm004_warren_2012_o.pdf

Warren CD (2011) DOE Progress report for lightweighting materials, 3. Polymer Composites, 3–15 to 3–41. Available at http://www1.eere.energy.gov/vehiclesandfuels/resources/vt_lm_fy11.html

# Chapter 6
# Recovery of Lignin

**Abstract** Recovery of lignin from different processes—kraft process, steam explosion process, organosolv process—is discussed in this chapter. Commercial suppliers of lignin are also listed.

**Keywords** Lignin · Lignin recovery · Kraft process · Steam explosion process · Organosolv process · Kraft lignin · Steam explosion lignin · Organosolv lignin

Several studies have investigated the possibility of using various technical lignins as raw material for carbon fibre production, e.g. lignosulfonates, organosolv lignins, steam explosion lignins and kraft lignins (Otani 1981; Kadla and Kubo 2002; Uraki et al. 1993; Sudo and Shimizu 1992; Sudo et al. 1993; Kubo and Kadla 2005; Attwenger 2014; Wang and Lu 2010; Stelte 2013). Organosolv lignin is generally much purer than the commercial kraft lignin (Huang 2009). Steam explosion and organosolv processes have been known in the pulping industry for decades and have a better environmental impact than kraft or sulphite processes.

By comparison with cellulose which is commercially used for pulp and paper production, lignin has very limited applications as a chemical and is not intentionally produced in industry. It generally originates in huge amounts as a side product when cellulose is isolated from the lignocellulosic material during the pulping process. There are various isolation technologies based on chemical or mechanical treatment for the separation of cellulose and hemicellulose from lignin (Smook 1992; Biermann 1996). Whereas mechanical treatments are used to mechanically separate the constituents from each other, the predominant chemical processes usually use harsh process conditions and pulping chemicals that uniquely alter the structure of lignin (Zakzeski et al. 2010). They either aim at the removal of cellulose and hemicellulose by solubilization leaving an insoluble lignin fraction behind or vice versa. Lignin from potential biorefineries that produce ethanol by fermentation of cellulosic sugars is derived from processes of the first category. The pulp and paper industry on the other hand uses chemical pulping processes resulting in an insoluble fibrous pulp and lignin-rich black liquor. Kraft pulping and to a little extent sulfite pulping are the predominant processes for wood in industry.

© The Author(s) 2017
P. Bajpai, *Carbon Fibre from Lignin*, SpringerBriefs in Materials,
DOI 10.1007/978-981-10-4229-4_6

Other pulping processes are so far mainly used in pilot-scale plants or in research (Lora 2008; Sixta 2006; Gosselink et al. 2004). The structure and properties of lignins vary based on different plant source they are separated from, such as hardwood, softwood, wheat straw or bamboo.

## 6.1  Kraft Lignin

The kraft process is also known as the sulphate process. It is the dominating chemical pulping technology worldwide with over 22 million tonnes produced in Europe as of 2011 (FAOSTAT 2012). The process is based on an alkaline solution of sodium hydroxide and sodium sulphide, which degrades the carbohydrates by alkaline and peeling hydrolysis. It accounts for 89% of the chemical pulp production. The high quality of the pulp including its high mechanical strength, the simplicity and rapidity of the process and also the technological advances have led to its predominant position (Gierer 1980). Kraft pulping gives high strength cellulose fibres and is used for various applications such as magazines, grocery bags and corrugated packaging (Sjöström 1993). A scheme of the kraft process in the paper industry including the main operation units is shown in Fig. 6.1. In the kraft pulping process, the chips are mixed with "white liquor" (solution of sodium hydroxide and sodium sulphide), heated to increase the reaction rate and then disintegrated into fibres by subjecting them to a sudden decrease in pressure. Typically some 150 kg of sodium hydroxide and 50 kg of sodium sulphide are

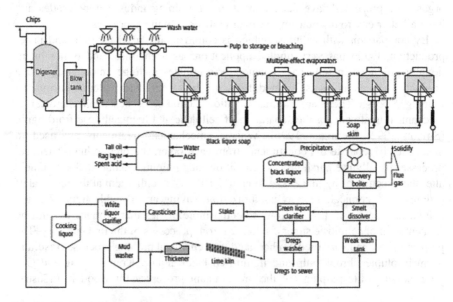

**Fig. 6.1** Kraft pulping and chemical recovery process. Based on Smook (1992)

required per tonne of dry wood. Like any chemical reaction, this process is, affected by temperature, time and concentration of chemical reactants. Temperature and time can be traded off against each other to a certain extent, but to obtain reasonable cooking times it is very important to have temperatures of about 150–165 °C, therefore pressure cookers are used. But, if the temperature is too high then the chips are unevenly delignified, so a balance should be obtained. The kinetics of the kraft pulping is understood well, but the reaction is heterogenous and therefore difficult to examine. To determine when to interrupt the cooking, a model relating temperature, time and cooking chemical charge is used. The degree of delignification is the most important parameter for determining pulp quality. It is usually expressed as Kappa number. This number is directly related to the amount of lignin still present in the cooked pulp. There are two different types of pulping systems—batch and continuous.

During this treatment, the lignin macromolecule is broken down by action of the hydroxide and hydrosulfide ions in the pulping liquor (white liquor) and results in smaller water/alkaline-soluble fragments (Chakar and Ragauskas 2004).

By adjusting the cooking time and temperature, the delignification of the fibre can be adjusted in order to control the final pulp quality. The strongly basic white liquor dissolves and partly degrades the lignin and hemicelluloses into the so-called black liquor. Besides lignin and hemicellulose, the black liquor contains low molecular mass compounds from the dissolution and partial degradation of wood constituents. These are carbohydrates, aliphatic acids and constituents such as extractives and inorganics, the latter also originating from the cooking chemicals.

The "black liquor" from the kraft treatment is in the next stages concentrated by evaporation and used as fuel in the recovery boilers (Vakkilainen 2000; Bajpai 2008). Therefore, Kraft pulp mills operate as highly integrated facilities that rely on lignin as an energy source. Since the present infrastructure of the process was optimized throughout the more than 130 years of operation, the recovery of lignin has not been practised broadly until recently. While the recovery boilers are essential to the environmental and economic performance of the kraft pulp mill, energetic improvements have led to the fact that current pulping processes are often limited by the energy load that can be handled therein (Öhman 2006). In this case, a capacity increase in pulp production is only possible when the thermal load to the recovery boiler is reduced. As modern kraft mills operate with an energy excess, a debottlenecking is possible in most of the present mills by partly removing the energy surplus from the black liquor in form of solid lignin (Olsson et al. 2006). Although this precipitated kraft lignin is often discussed as additional solid fuel in other areas of the pulp mill, , it represents a marketable product and an important raw material for chemical valorization (Lora 2008, Werhan 2013).

Different separation techniques to remove lignin from the kraft black liquor have been reported in the literature. The most common one is the precipitation of kraft lignin by acidification with carbon dioxide to a pH of 10–11 and subsequent filtration (Öhman 2006). Concentration and isolation by ultrafiltration is another approach (Jönsson et al. 2008).

One of the widely used processes for isolating large quantities of kraft lignin from black liquor is known as LignoBoost® developed by Innventia and Chalmers University of Technology (Chen 2014). This process has gained much attention and a pilot plant was successfully implemented in a Swedish pulp mill in 2007 (Tomani and Axegard 2007). In 2008, the technology was sold to Metso which together with Innventia developed the technology to industrial scale. LignoBoost makes it possible to extract lignin from the black liquor and to increase the liquor-burning capacity of chemical recovery boilers. The extracted lignin can be used to replace fossil fuel, or as a raw material in the chemical industry. LignoBoost gives pulp mills new potential to increase production, reduce costs and create new sources of income. It is capable of separating 25% of the lignin in the black liquor (corresponding to 24 t d − 1), thereby enabling 20–25% more pulp production (Ollila 2011). The first commercial Lignoboost plant will start its operation in the beginning of 2013. If this process continues succeeding, the amount of commercially available kraft lignin can be expected to increase and kraft lignin will get more and more important as a renewable raw material and feedstock.

Metso has patented the LignoBoost technology and has supplied the world's first commercial installation of LignoBoost technology to Domtar in North America (Christiansen 2013). The equipment has been intergrated with the Plymouth North Carolina pulp mill. The LignoBoost process separates and collects lignin from the pulping liquor. This is an important breakthrough for Metso's LignoBoost technology and provides several benefits to Plymouth North Carolina mill. Separation of a portion of the mill's total lignin production off-loads the recovery boiler and permits an increase in pulp production capacity. The recovered lignin will be used for several internal and external applications. This project is a potential game changer for the Pulp and Paper industry. It will allow pulp mills to have a new more profitable value stream from a product that was traditionally burned in the recovery boiler. Domtar Corporation is the largest integrated producer of uncoated freesheet paper in North America and the second largest in the world based on production capacity. It also produces paper-grade, fluff and specialty pulp. The Company manufactures, markets and distributes a wide range of, commercial printing and publishing, business and also converting and specialty papers.

Domtar's production of lignin started in February 2013 with a targeted rate of 75 tonnes a day (www.valmet.com). A wide range of applications and markets for BioChoice lignin are being developed. These include fuels, resins and thermoplastics. Having lignin available in large quantities and high quality from the Domtar plant will help develop the future lignin market for the industry.

Lignin available from the kraft pulping industry is very distinct from the original lignin found in plants. By action of the white liquor, the fragmentation of the lignin molecule in combination with the introduction of thiol groups from hydrosulfide anions render kraft lignin soluble. The fragmentation of lignin during kraft treatment mainly proceeds by cleavage of the $\alpha$-O-4 and $\beta$-O-4 linkages between the aromatic units. 80–85% thereof were found to be cleaved during kraft pulping of softwood resulting in an increase in phenolic hydroxyl groups (Toven and Gellerstedt 1999). All other interunit linkages in lignin including 4-O-5 aryl ether structures basically survive the harsh conditions in the kraft process (Gierer 1980). Apart from the fragmentation of the lignin molecule, condensation reactions which recombine lignin fragments also occur during kraft pulping. Thereby alkali-stable carbon-carbon linkages of the diaryl methane type are formed. Due to the high amount of stable carbon-carbon bonds which either remain unaffected or are even formed during kraft cooking, kraft lignin precipitated from black liquor represents a highly refractory material which is hard to depolymerize into aromatic units by chemical or biological processes. Analytical results of the molecular mass distribution usually report an average molar mass Mw of 2000–3000 $gmol^{-1}$ with polydispersities (PDI) from 2 to 3 for kraft lignin (Jacobs and Dahlman 2000). The functional group content and other characteristics (e.g. elemental composition, impurities) of kraft lignin have been well characterized (Mansouri and Salvadó 2006).

Out of the approximately 20 billion tonnes of lignin that are annually synthesized in nature, an estimated 70 million tonnes arise each year in kraft pulp mills during the production of pulp and regenerated fibres (Pye 2006). However, as kraft pulp mills have evolved as highly energetically integrated facilities, more than 99% of the so-called kraft lignin from the process is not recovered for industrial application but burned in the recovery boilers for the recovery of pulping chemicals and the provision of energy. According to Lora (2008), only about 60,000 tonnes of kraft lignin per year are commercialized, virtually all of it by MeadwestVaco in the United States. In combination with one million tonnes of lignosulfonates from the sulfite pulping industry and about 10,000 tonnes from the soda pulping industry, the total marketed amount of lignin is a little lower than 1.1 million tonnes annually. This accounts for less than 2% of the total amount of processed lignin (Gosselink et al. 2004). Production of other lignins which also use different pulping/isolation technologies is so far only conducted on small scale in pilot plants or in research and technical development. An overview of the main current lignin producers (production of $\geq 0.5$ kt/y) is presented in Table 6.1.

| Table 6.1 Suppliers of lignin | Borregaard (NO), TEMBEC (FR,US), Domsjö Fabriker (SE), La Rochette Venizel (FR), Nippon Paper (JP) | Lignosulfonates |
|---|---|---|
| | MeadwestVaco (US) | Kraft lignin |
| | GreenValue (USA) | Sodalignin |
| | Lignoboost/Metso (SE) | Kraft lignin |
| | Lignol (CA), Fraunhofer (DE) Dedini (BR) | Organosolv |
| | CIMV (FR) | Organosolv |
| | SEKAB (SE) | Hydrolysis lignin |
| | Inbicon (DK,US), Chemtex (IT,US,CN) | Hydrolysis lignin |

Based on Gosselink (2011)

## 6.2 Steam Explosion Lignin

In this process, high pressure steam applied on lignocellulosic material for a short period of time, followed by sudden explosion results in fibreization of the biomass (Ibrahim and Glasser 1999). Steam explosion was introduced and patented as a biomass pre-treatment process in 1926 by Mason (1926). The steam explosion process is a type of pre-treatment which enables easier fibre and has been shown to be a fundamental technology for biomass separation.

Patent by Mason (1926) describes a steam explosion process for the pre-treatment of wood. In this process, wood chips are fed from a bin through a screw loading valve in a masonite gun. The chips are then steam heated at a temperature of about 285 °C and a pressure of 3.5 MPa for about 2 min. The pressure is increased rapidly to about 7 MPa (70 bar) for about 5s, and the chips are then discharged through restricted orifices (slotted port) and explode at atmospheric pressure into a pulp. In general steam explosion is a process in which biomass is treated with hot steam (180–240 °C) under pressure (1–3.5 MPa) followed by an explosive decompression of the biomass that results in a rupture of the biomass fibres rigid structure. The sudden pressure release defibrillates the cellulose bundles, and this result in a better accessibility of the cellulose for enzymatic hydrolysis and fermentation. Depending on temperature and residence time, steam explosion can result in anything from small cracks in the wood structure, to total defibrillation of the wood fibres (Tanahashi 1990). Acetic acid is liberated from the wood, and this result in partial hydrolysis of the cell wall components (Glasser and Wright 1998). Use of diluted acids such as sulfuric or nitric acid can accelerate the process which result in higher hydrolysis rates of the hemicelluloses (Boussaid et al. 2000; Shevchenko et al. 2001; Bura et al. 2002).

General advantages of steam explosion processes compared to other pre-treatment technologies for chemical utilization of lignocellulose are presented in Table 6.2 (Garrote et al. 1999).

As a pre-treatment for microbial bioethanol or biogas production, steam explosion of biomass can be used as an environmental friendly pulping process (Cara et al. 2006). The application of steam explosion in biomass conversion, the techniques and their advantages were described by Wang and Lu (2010). Steam explosion requires less energy compared to many other methods and has low environmental impact, but because the lignin ($\sim$60%) is poorly solubilized and its structure is significantly broken down, steam explosion is not a suitable recovery process for further lignin processing so far (Hergert and Pye 1992).

| **Table 6.2** Advantages of steam explosion processes | No chemicals are used except water |
| --- | --- |
| | Good yield of hemicelluloses with low degraded byproducts |
| | Equipment corrosion is minimum due to a mild pH of reaction media when compared to acid hydrolysis processes |
| | Stages of acid handling and acid recycling are avoided— Disruption of the solid residues from bundles to individual fibres occur due to explosion effect |

## 6.3 Organosolv Lignin

Organosolv is based on heat treatment for biomass with an (aqueous) organic solvent at high temperatures. Commonly used solvents are presented below:

- Ethanol
- Methanol
- Acetone
- Organic acids (acetic acid and formic acid or combinations thereof).

Organosolv processes delignify lignocelluloses; the organic solvent function as lignin extractant, whereas the hemicellulose is depolymerized using acid-catalyzed hydrolysis. In general, organosolv processes aim to fractionate the lignocellulosic biomass as much as possible into its individual major fractions. This is in contrast to other pre-treatment technologies such as dilute acid hydrolysis and steam explosion. These technologies simply make the cellulose fraction suitable for further processing without recovery of a purified lignin fraction. Organosolv lignin has a high purity. It has limited amounts of residual carbohydrates and minerals due to isolation process. As a result, its application spectrum is broader in comparison to the more impure lignin-containing residues derived from conventional pre-treatments which are targeted basically towards the production of cellulose for paper or second generation bioethanol. The later are a complex mixture of unconverted carbohydrates, lignin, minerals and process chemicals or microbial residues. Hardly any applications for such complex byproducts have been identified other than combustion of combined heat and power. Organosolv lignins have a relatively low molecular weight having a narrow distribution and a very low sulphur content.

Organosolv pulping involves treating a lignocellulosic feedstock such as chipped wood or grasses with an organic solvent at temperatures ranging from 130–200 °C. A benefit of organosolv solvents is that they can be easily recovered by distillation. This leads to less water pollution and elimination of the odour usually associated with kraft pulping.

In 1998, Black et al. (1998) reported a method for separating lignocellulosic material into its three major components for further processing. They showed that lignin is present in the organic solvent and in the aqueous phase. But there are also certain disadvantages to this process that should be considered. Organic solvents are expensive and require a highly accurate processing environment as temperature and pressure ranges and increase the energy costs significantly. But investigations are ongoing to solve the problems and to open the way for this promising method. Hergert and Pye (1992) has given a detailed rationale for the organosolv process describing all relevant work since 1987. They have concluded that two of many organosolv pulping processes are most promising, the Alcell process for hardwoods and the Organocell process for softwoods. A modified alcohol pulping and recovery process using 50% ethanol at 195 °C called alcohol pulping and recovery process (APR) was further improved by the company Alcell Developments Inc. Organocell GmbH owns the Organocell process. This was the most advanced new process to be

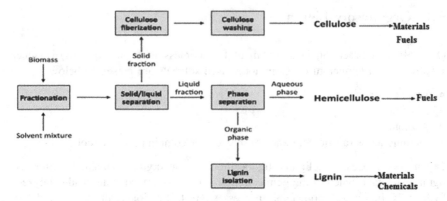

**Fig. 6.2** Organosolv fractionation method. Based on Baker and Rials (2013), Bozell et al. (2011)

implemented on an industrial scale in the year 1992. The lignin was recovered using an organosolv fractionation method developed at the National Renewable Energy Laboratory by Bozell et al. (2011). This method separates lignocellulosic biomass into its three main components: cellulose, hemicellulose and lignin (Fig. 6.2). In this process, an experimental reactor developed at the University of Tennessee's Centre for Renewable Carbon was used to perform the organosolv fractionation of the raw material. A mixture of MIBK, ethanol and water was used in the presence of an acid promoter. After the fractionation, a solid fraction and a liquid fraction are separated. The solid fraction contains the remaining cellulose. The liquid fraction is called black liquor and contains lignin and hemicellulose. The black liquor is then separated into an aqueous phase, containing mainly hemicellulose and an organic phase, containing mainly lignin.

Astner (2012) studied optimization of lignin yield over a range of process temperatures from 120–160 °C and a sulphuric acid concentration of 0.025–1 M. A maximum lignin yield of 81% at a fractionation time of 90 min at 160 °C using 0.1 M sulfuric acid concentration with a feedstock ratio of 90% switchgrass and 10% tulip poplar was obtained.

Another study on the organosolv fractionation process was conducted by Maraun (2013). The findings of this study describe solvent composition and run conditions duration in the presence of feedstock contamination. A maximum mean lignin yield of 85.7% could be obtained at a fractionation time of 56 min at a temperature of 160 °C using 0.1 M sulfuric acid and a feedstock mixture containing 90% tulip poplar and 10% switch grass.

The organosolv fractionation separates the lignin from the hemicelluloses and removes non-cellulose components. This pre-treatment also reduces the crystallinity of the cellulose and creates a specific surface area. Finally a pure, high-quality lignin is recovered useable for further processing steps like melt spinning. The properties of organosolv lignin differ from other technical lignin. The major features are low molecular weight and high chemical purity (Lora 2008).

# References

Astner FA (2012) Lignin yield maximization of lignocellulosic biomass by taguchi robust product design using organosolv fractionation. University of Tennessee, Knoxville, 137 p

Attwenger A (2014) Value-added lignin based carbon Fibre from organosolv Fractionation of poplar and Switchgrass. Thesis, Master of Science, University of Tennessee—Knoxville, USA

Bajpai P (2008) Chemical recovery in pulp and paper making. PIRA International, Leatherhead, 166 p

Baker DA, Rials TG (2013) Recent advances in low-cost carbon fibre manufacture from lignin. J Appl Polym Sci 130:713–728

Biermann CJ (1996) Handbook of pulping and papermaking, 2nd edn. Academic, San Diego, 754 p

Black SK, Hames BR, Myers MD (1998) A. A. patent 5,730,838. M. R. Institute

Bozell JJ, Black SK, Myers M, Cahill D, Miller WP, Park S (2011) Solvent fractionation of renewable woody feedstocks: Organosolv generation of biorefinery process streams for the production of biobased chemicals. Biomass Bioenergy 35:4197–4208

Boussaid A, Esteghlalian R, Gregg J, Lee KE, Saddler JN (2000) Steam pretreatment of Douglas Fir Wood Chips. Appl Biochem Biotechnol 84–86:693–705

Bura R, Mansfield SD, Saddler JN, Bothast JR (2002) $SO_2$ catalyzed steam explosion of corn fibre for ethanol production. Appl Biochem Biotechnol 98–100:59–72

Cara C, Ruiz E, Ballesteros I, Negro MJ, Castro E (2006) Enhanced enzymatic hydrolysis of olive tree wood by steam explosion and alkaline peroxide delignification. Process Biochem 41:423–429

Chakar FS, Ragauskas AJ (2004) Review of current and future softwood kraft lignin process chemistry. Ind Crops Prod 20(2):131–141

Chen MCW (2014) Commercial viability analysis of lignin based carbon fibre. Master thesis, Simon Fraser University

Christiansen (2013) Metso-supplied world's first commercial LignoBoost plant successfully starts up at Domtar in the USA http://www.metso.com/news/2013/4/metso-supplied-worlds-first-commercial-lignoboost-plant-successfully-starts-up-at-domtar-in-the-usa/. Press release 30 Apr 2013

FAOSTAT (2012) Forestry production and trade [Online]. Food and agriculture organization of the United Nations. Available http://faostat3.fao.org/home/index.html#VISUALIZE_BY_DOMAIN. Accessed 9 Nov 2012

Garrote G, Dominguez H, Parajo JC (1999) Hydrothermal processing of lignocellulosic materials. Holzforschung 57:191–202

Gierer J (1980) Chemical aspects of kraft pulping. Wood Sci Technol 14(4):241–266

Glasser WG, Wright RS (1998) Steam assisted biomass fractionation. II. Fractionation behavior of various biomass resources. Biomass Bioenergy 14:219–235

Gosselink RJA (2011) Lignin as a renewable aromatic resource for the chemical industry. Ph.D. Thesis—Proefschrift Wageningen, 2011

Gosselink RJA, de Jong E, Guran B, Abächerli A (2004) Coordination network for lignin–standardisation, production and applications adapted to market requirements (EUROLIGNIN). Ind Crops Prod 20(2):121–129

Hergert HL, Pye EK (1992) Recent history of organosolv pulping. Notes Tech, pp 9–26

Huang X (2009) Fabrication and properties of carbon fibres. Materials 2009, 2, 2369–2403. doi:10.3390/ma2042369

Ibrahim M, Glasser WG (1999) Steam-assisted biomass fractionation. part III: a quantitative evaluation of the "clean fractionation" concept. Bioresour Technol 70:181–192

Jacobs A, Dahlman O (2000) Absolute molar mass of lignins by size exclusion chromatography and MALDI-TOF mass spectroscopy. Nord Pulp Pap Res J 15(2):121–127

Jönsson AS, Nordin AK, Wallberg O (2008) Concentration and purification of lignin in hardwood kraft pulping liquor by ultrafiltration and nanofiltration. Chem Eng Res Des 86(11):1271–1280

Kadla JF, Kubo S, Gilbert RD, Venditti RA, Compere AL, Griffith WL (2002) Lignin-based carbon fibres for composite fibre applications. Carbon 40(15):2913–2920

Kubo S, Kadla JF (2005) Lignin-based carbon fibres: effect of synthetic polymer blending on fibre properties. J Polym Environ 13(2):97–105

Lora J (2008) Industrial commercial lignins: sources, properties and applications. In: Belgacem MN, Gandini A (eds) Monomers, polymers and composites from renewable resources. Elsevier, Amsterdam, pp 225–241

Maraun H (2013) Maximizing lignin yield using experimental design—analyzing the impact of solvent composition and feedstock particle size on the organosolv process in the presence of feedstock contamination. University of Tennessee, Knoxville, 124 p

Mansouri N-EEl, Salvadó J (2006) Structural characterization of technical lignins for the production of adhesives: application tolignosulfonate, kraft, soda-anthraquinone, organosolv and ethanol process lignins. Ind Crops Prod 24(1):8–16

Mason WH (1926) Process and apparatus for disintegration of wood and the like. US Patent: 1578609

Öhman F (2006) Precipitation and separation of lignin from kraft black liquor. Ph.D. thesis

Olsson MR, Axelsson E, Berntsson T (2006) Exporting lignin or power from heat-integrated kraft pulp mills: a techno-economic comparison using model mills. Nord Pulp Pap Res J 21(4):476–484

Ollila J (2011) Metso to deliver the first commercial LignoBoost plant to Domtar in North America. Press release on www.metso.com on December 15

Otani S (1981) Carbonaceous mesophase and carbon fibres. Mol Cryst Liq Cryst 63:249–264

Pye EK (2006) Industrial lignin production and applications. In: Kamm B, Gruber P, Kamm M (eds) Biorefineries-industrial processes and products, Wiley-VCH Verlag GmbH & Co. KGaA, Weinheim, pp 165–200

Shevchenko SM, Chang K, Dick DG, Gregg DJ, Saddler JN (2001) Structure and properties of lignin in softwoods after $SO_2$ catalyzed steam explosion and enzymatic hydrolysis. Cellul Chem Technol 35(5–6):487–502

Sjöström E (1993) Wood chemistry: fundamentals and application. Academic Press, Orlando, p 293

Sixta H (2006) Handbook of pulp. Wiley-VCH Verlag GmbH & Co. KGaA, Weinheim

Smook GA (1992) Handbook for pulp and paper technologists. Angus Wilde Publications, Inc, Vancouver, 425 p

Stelte W (2013) Steam explosion for biomass pre-treatment. Danish Technol Inst, Denmark

Sudo K, Shimizu K (1992) A new carbon-fibre from lignin. J Appl Polym Sci 44(1):127–134

Sudo K, Shimizu K, Nakashima N, Yokoyama A (1993) A new modification method of exploded lignin for the preparation of a carbon-fibre precursor. J Appl Polym Sci 48(8):1485–1491

Tanahashi M (1990) Characterization and degradation mechanisms of wood components by steam explosion and utilization of exploded wood. Wood Res 77:49–117

Tomani P, Axegard P (2007) The ILI umbrella programme and other existing and new approaches in lignin research. In ILI 8th Forum, The International Lignin Institute, pp 109–113

Toven K, Gellerstedt G (1999) Structural changes of softwood kraft lignin in oxygen delignification and prebleaching. In: 10th International Symposium on Wood and Pulping Chemistry, vol 2, pp 340–345

Uraki Y, Kubo S, Sano Y, Sasaya T, Ogawa M (1993) Melt Spinning of Organosolv Lignin. In: 7th International Symposium On Wood and Pulping Chemistry, Beijing, China

Vakkilainen EK (2000) Chapter 1: chemical recovery. In: Gullichsen J, Paulapuro H (eds) Papermaking science and technology, vol 6B. Fapet Oy, Helsinki, p 7

Wang XT, Lu LS (2010) Steam explosion pretreatment technique and application in biomass conversion. Adv Mater Res-Switz 113–114:525–528

Werhan H (2013) A process for the complete valorization of lignin into aromatic chemicals based on acidic oxidation. Degree of Doctor of Sciences, ETH Zurich

Zakzeski J, Bruijnincx PCA, Jongerius AL, Weckhuysen BM (2010) The catalytic valorization of lignin for the production of renewable chemicals. Chem Rev 110(6):3552–3599

# Chapter 7
# Lignin as a Precursor for Carbon Fibre Production

**Abstract** The available information on production, properties of carbon fibres from lignin is presented in this chapter.

**Keywords** Lignin · Carbon fibre · Production · Properties

Lignin is potentially a new precursor material for the production of carbon fibre and shows several advantages over PAN and pitch for the production of commercial carbon fibre (Norgren and Edlund 2014; Baker and Rials 2013; Bajpai 2013). Lignin is relatively inexpensive, readily available and structurally rich in phenyl propane group having high carbon content (60%). A significant source of lignin in the form of black liquor results from pulping of both wood and non-woody materials by using the Kraft process. Presently, black liquor is burned for its fuel value and has not been used commercially to produce carbon fibre. Currently, the hemicelluloses extraction process at near neutral condition is being developed to convert hemicelluloses into ethanol and acetic acid, but the extracted lignin is underutilized. This process presents an opportunity to obtain a clean lignin feedback suitable for the manufacture of carbon fibre. The economic viability of this process could possibly be improved provided lignin could be recovered as a by-product. It also presents an opportunity for the traditional pulping industry to generate an additional source of revenue (Attwenger 2014).

Several researchers have investigated the production of carbon fibres from lignin (Zhang 2016; Sudo and Shimizu 1987, 1989, 1992, 1994; Sudo et al. 1988, 1993; Luo 2004; Luo et al. 2011; Norberg 2012; Norberg et al. 2012; Nordström 2012; Nordström et al. 2012; Kubo et al. 1997, 1998; Kadla et al. 2002a, b; Baker 2011; Kadla and Kubo 2005; Kubo and Kadla 2005a, b; Ito and Shigemoto 1989; Uraki et al. 1995, 1997, 2001; Uraki and Kubo 2006; Eckert and Abdullah 2008). Extensive work in this area has been conducted by Dr. Bakers group (Baker 2010a, b, 2011; Baker et al. 1969, 2005, 2008a, b, 2009a, b, 2010a, b, c, 2011, 2012, 2013). *Excellent review on this topic has been published by* Baker and Rials (2013) and Frank et al. (2012, 2014). According to Baker and Rials (2013) lignin used to manufacture carbon fibre of higher strengths should not only be of high purity, but it should also

© The Author(s) 2017                                                                                     43
P. Bajpai, *Carbon Fibre from Lignin*, SpringerBriefs in Materials,
DOI 10.1007/978-981-10-4229-4_7

have a narrow molecular weight distribution; The difference between Tg and Ts should be small.

The studies have mainly focused towards improving the melt-spinning of lignin fibre and their conversion to carbon fibre with reduced cost. Carbon fibre from lignins without using any additives or chemical modifications applied before extrusion; Carbon fibre from lignins that were either coextruded, or chemically modified before extrusion and submicron carbon fibres from lignin have been investigated.

## 7.1 Production from Different Types of Lignin

The development of an alternative precursor for carbon fibre which was based on a renewable material such as lignin, was identified a long time ago. Otani et al. (1966, 1969) described various methods of producing fibre from different types of lignin such as softwood kraft lignin, hardwood kraft lignin, and alkali softwood lignins using the melt-spinning and dry spinning methods, and their conversion to carbon fibre, graphite fibre, and activated carbon fibre. The strength of carbon fibre produced from melt-spun precursor ranged up to 800 MPa. In the examples for dry-spinning, water or sodium hydroxide aqueous solutions were used as solvent for both and thiolignin. Carbon fibre produced from this dry-spinning process possessed strength that ranged only up to 300 MPa. Otani (1967) reported methods of carbon fibre production from different precursors, including lignin, and also the physical properties required for different types of industrial applications.

Kayocarbon fibre, a commercial carbon fibre from lignin, was produced by Nippon Kayaku Co in Japan in the 1960s (Fukuoka 1969). The poor mechanical properties of the carbon fibre product forced this project to be abandoned. In this process, the fibre was produced from lignosulfonate, using polyvinyl alcohol added as the plasticizer, and then dry spun (Donnet and Bansal 1990; Minus and Kumar 2005).

New attempts were made in the 1990s, and several different types of lignins—Steam explosion lignin, Organosolv lignin, and Kraft lignin were thoroughly studied as precursors for Carbon fibre (Sudo and Shimizu 1992, 1994; Sudo et al. 1993; Peebles 1994; Uraki et al. 1995, 1997, 2001; Uraki and Kubo 2006; Kubo et al. 1998; Kadla et al. 2002a, b; Baker 2011; Kubo and Kadla 2005a, b; Braun et al. 2005; Luo 2004; Schmidt et al. 1995).

Kayacarbon left the market since the petroleum-based precursors have been developed more rapidly which gives higher strength at reduced price. The lower cost melt spinning route was not used, but this may have been due to difficulties in transitioning from small-scale single fibre extrusion equipment to larger multifilament facilities with substantially longer residence times, which cause the lignin to polymerize during extrusion.

Mansmann (1974) reported methods for the co-spinning of lignosulfonates with polyethylene oxide with rapid conversion to carbon fibre, providing fibres with strengths of up to 0.834 GPa. Few examples were provided in which salts were added for improving the conversion kinetics.

Gould (1974) reported a method of producing improved carbon fibre by extracting pulp waste lignin with an alcohol, followed by demethylation, and polymerization by using irradiation or thermal treatment before conversion.

Sudo and Shimizu (1987) and Sudo et al. (1988) reported the preparation of lignins using high pressure steam treatment of wood with further treatment with organic solvent or alkali. The resulting lignins were first hydrocracked for reducing the molecular weight and complexity, and then treated to increase molecular weight to provide a precursor for melt-spinning. Conversion of the lignin fibres to carbon fibres showed strengths in the range of 30–80 kg/mm$^2$ and of 10–40 μm diameter.

Preparation of lignin was reported by Sudo and Shimizu (1989) whereby treatment with phenol in the presence of p-toluene sulphonic acid produced a melt spinnable product. Conversion of the melt spun fibre produced carbon fibres having strengths in the region of 52.8 kg-f/mm$^2$.

Ito and Shigemoto (1989) extracted the wood chips at high temperature with a mixture of water and cresols. The solvents were recovered by evaporation and the lignins were melt-spun to give thermoplastic fibre which could be rendered infusible at 3 °C/min to 200 °C. The lignins were found suitable for the production of carbon fibre.

Another lignin preparation method was reported by Ichikawa et al. (1992) in which lignins were phenylated but were then coextruded with various amounts of pitch for improving the properties of carbon fibre.

Sudo and Shimizu (1992) developed a process for the production of carbon fibre from hardwood lignin in which steam-explosion technology was used to isolate lignin from birch wood. Lignin was modified by using hydrogenation technique for improving the melt spinning. Chloroform and carbon disulfide were used for dissolving and separating the insoluble lignin fraction. The purified lignin was heated between 300 and 350 °C under vacuum for 30 min. This produced a molten viscous lignin having a softening point of 110 °C. It melted completely at 145 °C. This material was suitable for the preparation of fine filaments. Infusible lignin fibres were produced by thermostabilization of the filaments at 210 °C. The filaments were carbonized by heating from room temperature to 1000 °C at a heating rate of 5 °C/min in nitrogen.

Sudo et al. (1993) made a comparison of the chemical structure of the precursor with the crude lignin and concluded that there was a substantial elimination of aliphatic functional groups compared to the original starting material. These researchers investigated the use of phenolated lignin as a carbon fibre precursor. This work was conducted as an alternative to the use of hydrogenated lignin because of the high cost associated with producing hydrogen. Similar to the hydrogenation process, steam-exploded lignin was the crude feedstock used in the phenolysis process. The phenolysis reaction was performed by treating equal weights of phenol and crude lignin under vacuum at 180–300 °C for 2–5 h.

Para-toluene-sulfuric acid was used as the catalyst in the reaction. The resulting lignin-pitch was readily spun into fine filaments. The fine filaments were converted into carbon fibre after thermostabilization and carbonization. The overall yield for the process was 43.7% based upon the phenolated lignin. Properties of the carbon fibre from phenolated lignin are presented in Table 7.1. The tensile strength of the carbon fibre was approximately 455 MPa, but was not as high as that of the hydrogenated lignin.

Organosolv lignin has been used to prepare carbon fibre. The solvents used in the pulping processes used to liberate the crude lignin are acetic acid and ethanol. Crude lignin was obtained by acetic acid pulping of birch wood and used as a precursor for producing carbon fibre (Uraki et al. 1995). The organsolv lignin obtained by aqueous acetic acid pulping was used. The poly-dispersity of the resulting lignin and partial acetylation of hydroxyl groups during the pulping process was thought to be responsible for the ability of the crude lignin to be readily spun into lignin fibre due to a limited acetylation. Conversion of the melt spun fibres to carbon fibres with strengths having up to 0.355 GPa and moduli of 39.1 GPa were reported (Uraki et al. 1995, 1997). The conversion of these carbon fibres to activated carbon fibre was examined (Uraki et al. 1997). Table 7.1 shows data on the physical properties of carbon fibre produced from crude lignin obtained by acetic acid pulping of hardwood.

Uraki et al. (2001) studied the preparation of carbon fibre from a lignin recovered by the acetic acid catalyzed fractionation of Todo Fir. Solvent was used for the extraction of lignin and melt-spun on single filament apparatus. Single fibres could be oxidatively stabilized at rates of up to 3.0 °C/min to 250 °C, and carbonized at 180 °C/min to 1000 °C to give carbon fibres with strengths of 0.15 GPa. Also direct carbonization was reported. Increased efficacy compared with conventional carbon fibres was observed.

Kadla et al. (2002a, b) and Baker (2011) made a comparison of the Alcell lignin to lignin obtained from hardwood Kraft lignin and Indulin AT (softwood Kraft lignin). The Alcell lignin used by Kadla was obtained from Repap Enterprises in Newcastle, New Brunswick. These authors reported that the Indulin AT could not be spun into lignin fibre because of charring before melting. It was possible to melt spun both the Alcell and hardwood Kraft lignin into lignin fibre, but the Alcell

**Table 7.1** Properties of carbon fibre from different hardwood lignin

| Lignin type | Diameter (μm) | Elongation (%) | Tensile strength (Mpa) | Modulus of elasticity (GPa) |
|---|---|---|---|---|
| Phenolated hardwood lignin | NA | 1.4 | 455 | 32.5 |
| Hardwood lignin from acetic acid pulping | 14 ± 1.0 | 0.98 ± 0.25 | 355 ± 53 | 39.1 ± 13.3 |
| Hardwood alcell lignin | 31 ± 3 | 1.00 ± 0.23 | 388 ± 123 | 40.0 ± 14 |

Based on Sudo et al. (1993), Uraki et al. (1995), Kadla et al. (2002a, b)

lignin had a significantly lower spinning temperature; about 140 °C for Alcell lignin compared to 200 °C for hardwood Kraft lignin. Infusible lignin fibre could be produced during the thermostabilization process using Alcell lignin fibre provided the heating rate was maintained below 12 °C/hour. The physical properties of carbon fibre produced from Alcell lignin are presented in Table 7.1. The lignins were desalted before any measurements and/or extrusion and the hardwood-based samples were treated at 145 °C for one hour for removing potential volatiles. It was observed that that the hardwood lignins were readily melt spun into fibre but the softwood lignin had low thermal stability and therefore crosslinked during extrusion Nordström et al. 2013a. This was in agreement with the findings of Kubo et al. (1997). Alcell lignin with a $T_{g(onset)}$ of 68.2 °C could be single-filament spun at between 138–165 °C to produce lignin fibres which were oxidatively thermostabilized at a rate of 0.2 °C/min to 250 °C. Later carbonization at a rate of 3 °C/min to 1000 °C, produced fibres having strength of 0.388 GPa, 40 GPa modulus and 1.00% extensibility and 41.8% yield based on the starting Alcell fibres. Similarly, the hardwood kraft lignin with a $T_{g(onset)}$ of 83.3 °C was spun at 195–228 °C, providing fibres that were oxidatively thermostabilized at a rate of 2.0 °C/min under similar conditions. Carbonization of the thermoset fibres produced carbon fibres with 0.422 GPa strength, 40 GPa modulus and 1.12% extensibility and 48.1% yield.

Baker et al. (2008b, 2009b) reported the properties of two lignins which were hardwood kraft lignin and an organic purified derivative of the same. The kraft lignin was not readily melt-spinable but the organic purified lignin was, and continuous lignin tows could be spun to lower diameters (10 μm) and with high speed. But, conversion of the lignin fibre tow progressed slowly with oxidative thermostabilization rates as low as 0.01 °C/min being required before carbonization was possible. The resulting carbon fibres were of low yield, 32%, and had a lower strength of 0.517 GPa. The use of a thermal pre-treatment to increase $T_g$ and $T_s$ in case of Alcell™ was used to study resulting changes in the conversion kinetics of the lignin fibre into carbon fibre. The tensile properties and yield of the carbon fibres were improved by using these treatments, giving 0.710 GPa and 41%, respectively as compared to the original Alcell based carbon fibres, which showed a strength of 0.338 GPa and a yield of 31%. The multifilament tow processing time was reduced substantially from 14 days to 14 h, thereby establishing a firm link between lignin fibre $T_g$ and permissible oxidative thermostabilization rates (Baker et al. 2009). The lignin based carbon fibres produced highly graphitic structures when treated at temperatures higher than 2100 °C (Baker et al. 2009b, 2010a, b, c).

Baker et al. (2010a, b) reported that the use of a thermal pretreatment for adjusting the $T_g$ and melt flow properties of a lignin starting material would be more effectively used using a lignin having a narrower molecular weight distribution. Study was conducted with OP86, which had more favorable molecular weight properties compared to Alcell due to the method of manufacture. The thermal pre-treatment of OP86 gave a selection of lignins with differing $T_g$ and $T_s$ properties, which were examined for their multifilament melt-spinning properties. The resulting lignin fibre tows were studied for their carbonization and oxidative thermostabilization properties. The best carbon fibres produced showed tensile strengths of 1.07 GPa compared with the original lignin-derived carbon fibres having strength of 0.517 GPa; and with carbon yields, 55%, compared with the original lignin with 33% yield (Baker et al. 2008b). But, the oxidative thermostabilization times for these particular fibre tows was long (around 6–12 h), although some higher $T_g$ fibre samples could be treated in as little as 13 min, with oxidative thermostabilization from 25 to 250 °C at 20 °C/min. But, these latter fibres had much reduced tensile properties due to their relatively poor melt-spinning performance. Optimum levels of addition were found to give continuously spun multi-filaments with high conversion times. Other assessments using both the single filament and multi-filament spinning of lignin mixtures, such as PE, PET, PE, MPP, PAN and various other plasticizers were also conducted. But no improvement was observed in the mechanical properties of the resulting carbon fibres.

Baker et al. (2010a) melt spun lignins to give fibres of differing diameter and converted them to carbon fibre using identical conditions. It was found that while the cross-sectional area was doubled, strengths were reduced by only 15% and the modulus by 8% showing that although there is a diameter diffusion dependence, it is much reduced compared with other precursors (Tagawa and Miyata 1997). This is because lignin is already highly oxidized.

Eckert and Abdullah (2008) developed a method for acetylating softwood Kraft lignin for use in melt spinning lignin fibre. The acetylation process was performed by using acetyl chloride, acetic anhydride and acetic acid. The acetylation reaction was conducted between the temperatures of 70 and 100 °C with and without the use of a catalyst. The preferred catalysts for obtaining lignin acetate that could be readily melted during spinning were organic amines, particularly tertiary amines such as tri-ethyl amine, tri-methyl amine and pyridine. The temperature for the reaction was approximately 50 °C. The acetylation technique allowed softwood Kraft lignin fibre to be spun at diameter between 5 and 100 µm. But no data were presented for the overall process yield for the carbon fibre product and the physical properties.

Not much information is available regarding the effect of oxidative thermostabilization on lignin $T_g$. The chemical and polymeric changes which occur during thermal conversion were reported by Braun et al. (2005). Homogenized hardwood kraft lignin strands were extruded ground and directly heat-treated at different rates and temperatures (Kadla et al. 2002a, b; Baker 2011). Changes in functional groups, elemental composition, thermal properties and molecular weight were studied. These researchers found that the oxidative thermostabilization process was consistent with autocatalysis. Increase in oxygen content occurred at temperatures of up to 250 °C, followed by reducing contents with high temperatures. It was found that for avoiding softening during heat treatments, the oxidative thermostabilization of lignin would have to take place using rates of less than 0.06 °C/min. This was much lower rate than had been used earlier (2 °C/min) for the same lignin when spun and thermally treated as a single filament (Kadla et al. 2002a, b; Baker 2011) but was consistent with the findings of Baker et al. (2012) using a very similar lignin when multifilament extrusion was used to produce the fibres.

Researchers at Innventia AB have explored the manufacture of carbon fibre from lignins obtained from their Lignoboost process (Gellerstedt et al. 2010; Tomani 2010). The process has been designed to work along with kraft pulping processes to recover lignins from a portion of the lignin-rich stream which would normally be incinerated in a recovery boiler for recovering chemicals and energy. In this process, the industrial black liquor is treated with carbon dioxide to precipite lignin, which washed with dilute acid and recovered. So the resulting lignin is relatively free of salts and carbohydrates compared to the kraft lignins (Norberg 2012). Ultrafiltration of the black liquor, before isolation, resulted in a kraft lignin of a high degree of purity. The type of kraft lignin used governed the lignin's response to thermal treatment. The lignins were rendered more stable by oxidative stabilization, and there was a 10–20% increase in the final yield after carbonization, compared with stabilization without oxygen. The products that were obtained suggested that radical, oxidation, condensation and rearrangement reactions were the main reactions occurring during oxidative stabilization. Due to structural differences between kraft lignins from softwoods and from hardwoods, it was possible to carry out thermal stabilization in an inert atmosphere using only heat for the softwood kraft lignin fibres. A one-step operation was all that was required to perform stabilization and carbonization on softwood kraft lignin fibres, suggesting no need for a separate stabilization step with these fibres, which may reduce processing costs.

Nordstrom et al. (2012) reported the properties and processing of four lignins produced from both softwood and hardwood kraft black liquors using the Lignoboost process, both with and without prior treatment by ultrafiltration for reducing polydispersity. Each lignin was evaluated for melt-spinning capability. It was found that the permeate lignins, could be spun into continuous filaments whereas the non-permeate could not easily be spun and SKL could not be melt spun into fibre. Comparison of the molecular weight data for SKLP and SKL showed that though their $T_g$s were same, the molecular weight was almost double for the latter showing the presence of significant amount of infusible components in the lignins which were solubilized when acetylated for SEC measurement; and this was seen in the reduced magnitude $T_g$ of SKL relative to the permeate lignins. Information regarding the use of low $T_g$ permeate lignins for plasticizing high $T_g$ parent lignins, and the resulting effects observed in $T_g$ measurements and improved fibre spinning were provided.

Lignin recovered from the near-neutral hemicellulose extraction process was studied as a precursor suitable for production of carbon fibre by Luo et al. (2011). Crude lignin was precipitated from the wood extract by using sulfuric acid to reduce the pH value to 1.0. The crude lignin extract was improved by using hydrolysis to break the lignin-carbohydrate bonds and to remove carbohydrates which contaminate the lignin. The solids were precipitated which were separated by filtration, washed with water, and then dried. Lignin was recovered by using the hydrolysis method. It was found to be high in carbon and total lignin, low in inorganic contamination and insoluble material, but high in volatile material. The recovered lignin was found to be thermally spun into lignin fibres, but the spun fibres were brittle, because of its low-molecular weight and the glassy nature of lignin.

Micrographs obtained using scanning electron microscopy showed imperfections on the surface and in the interior microstructure of the carbon fibre when compared with micrographs taken of commercial carbon fibre manufactured using PAN and pitch. These imperfections were possibly related to the high volatile material content in the samples and the slow heating rate during the carbonization process. Baker (2011) has suggested that the increasing rate of heating during carbonization can reduce the degree of brittleness and improve mechanical properties.

Norberg et al. (2012) investigated selected grades of softwood and hardwood kraft lignin. Both grades were further fractionated with ultrafiltration of the black liquors to generate permeate lignins. This particular grade of fractionated softwood kraft lignin was successfully melt-spun into precursor fibres. Also, 10% of permeate hardwood kraft lignin was added into the unfractionated softwood kraft lignin as a softening agent to achieve melt-spinning. The advantage of this process is the fast stabilization achieved in 85 min, which is faster than most of the lignin fibre stabilization reported in the literature. However, the carbonized fibres had large diameters (above 50 μm) and the carbon fibre strength was not reported. The large cross-section area of those carbon fibres would lead to poor tensile strength. In another study reported by Nordström et al. (2013b), using the same separation method, the permeate hardwood kraft lignin was blended with different unpermeated softwood kraft lignin with various ratios. They found that the suitable spinning temperature of the mixtures decreased as permeate hardwood fraction increased. The tensile strength reported for those lignin based carbon fibres were all below 400 MPa. Several studies have made attempts to introduce acrylonitrile or polyacrylonitrile into lignin to develop carbon fibres.

Xia et al. (2016) prepared a lignosulfonate-AN copolymer which is also soluble in DMSO and insoluble in water. They conducted wet-spinning using lignosulfonate/PAN mixture and lignosulfonate-AN copolymer, respectively. Macrovoids were found in the as-spun fibres derived from lignosulfonate/PAN mixture, acting as defect in the resulting carbon fibres. But the aspun fibres obtained from lignosulfonate-AN copolymer were solid without voids. The resulting carbon fibres from this copolymer showed a strength of 540 MPa. Similarly, lignosulfonate/PAN blend solution was prepared for wet spinning (Dong et al. 2015). The lignin content used in this study ranged from 15 to 47%. Voids were still observed in the as spun fibres.

Brodin et al. (2012) also reported the oxidative stabilization of kraft lignins using powders, including spruce, eucalyptus and birch, and used an experimental design to determine the effects of various parameters on stabilization and carbonization yields. The input factors were feedstock, isothermal time, oxidative thermostabilization temperature and heating rate; the responses were carbonization yield, oxidative stabilization yield, and $T_g$. They showed that the carbon yield of lignin increases with the severeness of oxidative thermostabilization conditions used for both eucalyptus and birch, but not for pine lignin which showed an optimum condition. Under the conditions used for stabilization, almost all of the lignin powders showed a $T_g$ and thus few were completely thermostabilized during treatment. In contrast to this study, Baker et al. (2010a, b, c) showed that there was an optimum rate of oxidative thermostabilization for improving carbon fibre yields, below which carbon yield was found to be reduced as the balance of oxidation versus degradation became in favor of the latter, and above which lignin volatilization predominated over oxidation and crosslinking. However, Brodin et al. (2012) provided XPS data and suggested a mechanism by which lignin oxidative thermostabilization occurred, and showed differences in lignin functionality across the fibre diameter which showed air diffusion dependence. This is similar in concept to the accepted sheath-core differences in structure observed for both polyacry-lonitrile and mesophase pitch carbon fibres which cause the interior of fibre cross sections to be of lower strength.

Lignins obtained from other delignification processes have also been suggested as precursors for carbon fibre. Luo et al. (2011) described the recovery of lignins from the green liquor process, which has become a popular process directed towards biorefinery implementation. Lin et al. (2012) reported the production of carbon fibre from lignins obtained by the polyethylene glycol solvolysis of wood (Kurimoto et al. 1999) which produced carbon fibre with diameters of about 10.2 μm and tensile strengths of up to 0.457 GPa.

Prauchner et al. (2005) studied the use of a lignin-rich pitch precursor obtained from the slow pyrolysis of eucalyptus used to produce charcoal in Brazil. Condensation of the volatiles generated during treatment produced a tar which was distilled to yield a crude pitch residue. Thermal treatment of the lignin pitch increased its softening temperature from 76.1 to 134 °C which could be melt spun at 175–180 °C to produce single filaments having around 40 μm diameter. Oxidative thermostabilization of the filaments proceeded at 0.08 °C/min to 180 °C and carbonization at 2.0 °C/min to 1000 °C, producing carbon fibre of 0.129 GPa average strength. Qiao et al. (2005) used lignin-rich mixed hardwood tars and bamboo tars. These tars were polymerized using an aqueous solution of formaldehyde catalyzed by oxalic acid and to increase softening temperatures. The resulting pitches were melt spun at 130 °C and 180 °C for those obtained from hardwood and bamboo, respectively. Then the fibres were rendered infusible by treating at 200 °C at up to 0.3 °C/min, and carbonized at rates of up to 2 °C/min to 1000 °C. This resulted in carbon fibres with diameters as low as 13.4 μm, tensile strengths of up to 632 MPa and moduli of 44 GPa.

Qin and Kadla (2011) used organically modified montmorillonite into a pyrol-ysis lignin as a reinforcement. At organoclay loading below 1.0 wt%, tensile strength of resulting carbon fibres was improved by 12%. The resulting carbon fibres had a large diameter of more than 45 μm, and the tensile strength was below 500 MPa.

Qin and Kadla (2012) reported a process for recovering pyrolytic lignin from bio-oil which was obtained as a product of the pyrolysis of mixed hardwood sawdust (Scholze and Meier 2001). The pyrolytic lignin had a $T_g$ of 70 °C and was thermally treated under different conditions to give lignins with $T_g$s of up to 111 °C. These researchers presented the effects of the treatments on pyrolytic lignin in terms of elemental composition, NMR, molecular weight, oxidative thermostabilization response, fibre spinning performance and finally, carbon fibre strength. The lignin with the best melt-spinning performance was selected for conversion giving 49 μm diameter carbon fibres with an average strength of 0.370 GPa and a modulus of 36 GPa; the yield with reference to the spun fibre was high, 52%. The pyrolytic lignin-based carbon fibres were found comparable with those prepared from the hardwood lignin and Alcell lignin used for comparison.

Several patents have been applied/issued in the last few years:

Yang et al. (2011) obtained a patent for manufacturing carbon fibres using lignin based copolymer, based on lignin mixtures containing two or more of softwood lignin, hardwood lignin and pitch.

Patent was granted to Berlin (2011) which provided derivatives of native lignin having a certain carbon content and/or a certain alkoxy content and their use in carbon fibres. Increased alkoxy contents and/or carbon contents actually provide improved lignin-based carbon fibres. The particular properties of the so-called derivatives of native lignin do not appear to be particular to the fractionation process used, and so apply to any lignin that could be produced. The carbon content of a lignin for carbon fibre production should be greater than 64.5%, and/or the alkoxy content greater than 0.35 mmol/g for softwood lignins; 0.45 mmol/g for hardwood derived lignins, and 0.25 mmol/g for lignins derived from annual fibre.

Kim et al. (2011) reported a method for manufacturing lignin with high melt processibility.

A method was patented by Wohlmann et al. (2012) for lignins having a $T_g$ in the range 90–160 °C, a polydispersity of less than 28, an ash content of less than 1% (wt), and a volatiles content of less than 1% (wt).

Sjoholm et al. (2012) reported methods for producing fractionated softwood and hardwood alkali lignins, their mixtures and mixtures with parent lignins, and the extrusion properties of the materials produced.

Kadla et al. (2002a, b) and Baker (2011) studied the use of lignin-polyethylene oxide (PEO) blends as precursors for the production of carbon fibre. They used commercially available Kraft hardwood lignin without chemical modification to produce carbon fibres by thermal spinning followed by carbonization. Fibre spinning was made possible by the addition of the PEO to the commercial hardwood lignin. The lignin-PEO blends could be converted into fibre by the addition of 3–5% PEO. Beyond 5% PEO addition, fibre fusing occurred during thermal stabilization. The physical properties of carbon fibre from lignin-PEO blends are presented in Table 7.2.

Although the addition of PEO into the blend improved fibre spinning, the physical properties of the carbon fibre were not improved. Kubo and Kadla (2005) discussed the applications of a variety of lignin–synthetic polymer blends as precursors materials for the production of carbon fibre. Fibre spun from both unmodified hardwood Kraft and organosolv lignin proved to be brittle and difficult to handle. These researchers solved this problem by using blends of lignin and synthetic polymers of poly (ethylene terephthalate), polyethylene oxide and polypropylene to reduce the brittleness and improve the physical properties of the spun fibre properties. Blends were prepared that incorporated up to 25% of the synthetic polymer.

**Table 7.2** Physical properties of carbon fibres from blends of lignin-PEO

| Source | Fibre diameter (μm) | Elongation (%) | Tensile strength (MPa) | Modulus of elasticity (GPa) |
|---|---|---|---|---|
| HardWood Lignin | 46 ± 8 | 1.12 ± 0.22 | 422 ± 80 | 40 ± 11 |
| Lignin-PEO (97–3) | 34 ± 4 | 0.92 ± 0.21 | 448 ± 70 | 51 ± 13 |
| Lignin-PEO (95–5) | 46 ± 3 | 1.06 ± 0.14 | 396 ± 47 | 38 ± 5 |

Based on Kadla et al. (2002a, b)

The physical properties of the lignin-based polymers were found to be dependent upon three factors: (1) the source and properties of the lignin, (2) the amount and physical properties of the synthetic polymer being incorporated, and (3) chemical interactions between the components. Incorporating synthetic polymers into the precursor blends enhance the ability of the lignin-polymer blend to be spun into fibre, reduced the brittleness of the spun fibre, and improved its flexibility. The physical properties of some lignin-polymer blends are summarized in Table 7.3.

Adding polyethylene terephthalate to lignin to form a precursor blend (Table 7.3) raised both the modulus of elasticity and the tensile strength of the final carbon fibre product. Adding polypropylene to lignin-polymer blend precursor did not improve the physical properties. Kubo and Kadla (2005) imply that adding polyethylene oxide to lignin is a desirable blend as a precursor for producing lignin fibres, although the properties of the final carbon fibres were not given in the paper.

Kubo and Kadla (2004, 2005a) reported the effects of kraft lignin structure on miscibility of PEO and the relationship between structure, thermal properties and organosolv lignin/PEO blend behavior. They also reported the use of polypropylene instead of PEO, and despite the fact that polypropylene was not much miscible with the lignin, it was used to aid in the manufacture of hollow core carbon fibres, and afterwards, carbon fibres with high macroporosity but low mesoporosity (Kadla et al. 2002a, b; Baker 2011; Kubo and Kadla 2005a).

Kubo and Kadla (2005a) further reported the effects of blending a hardwood kraft lignin with polyethylene terephthalate and various polypropylenes on efficacy of fibre spinning and the thermal properties of fibre. They report that oxidative

**Table 7.3** Physical properties of carbon fibres from blends of lignin

| Source of lignin | Fibre diameter (μm) | Elongation (%) | Tensile strength (MPa) | Modulus of elasticity (GPa) |
|---|---|---|---|---|
| Lignin | NA | 1.07 | 422 | 39.6 |
| Lignin-PET (75–25) | NA | 0.77 | 511 | 66.3 |
| Lignin-PP (75–25) | NA | 0.50 | 113 | 22.8 |

Kubo and Kadla (2005b)

thermostabilization of a 75/25 w/w lignin/PET blend proceeded more rapidly (2 °C/min to 250 °C) than lignin fibres alone (0.2 °C/min to 250 °C). The mechanical properties of the carbon fibres produced from the blends were better in comparison with neat lignin analogues. The maximum single fibre strengths evaluated were 0.605 GPa with a modulus of 61 GPa for 100% lignin; and 0.703 GPa, with a modulus of 94 GPa for the 75/25 blend. The patent literature shows a interest in carbon fibre from lignin/polymer blends (Shen et al. 2007; Takanori et al. 2010).

Blends of lignin and PAN and their thermal conversion chemistry have been reported by several researchers (Bissett and Herriott 2012; Sazanov et al. 2007, 2008; Seydibeyoglu 2012; Lehmann et al. 2012a, b).

In the recent years, there has been a lot of interest in producing carbon fibre from lignin nanocomposite fibres. Sevastyanova et al. (2010) reported the inclusion of two types of modified montmorillonite organoclays into an organosolv lignin, and the extrusion properties of the composite materials at varying levels of inclusion. They found that the addition of the materials improved the effectiveness of fibre spinning and that the strength of the resulting lignin fibre increased by almost two times.

Qin and Kadla (2011) used a pyrolytic lignin which was isolated from bio-oil, and thermally treated to increase its molecular weight and glass transition temperature before compounding and fibre spinning (Scholze and Meier 2001). X-ray diffraction data revealed that the clays were well intercalated. In this case, the use of the organoclays did not significantly improve the strengths of the lignin fibres as they did for the organosolv lignin. The optimum level of addition was around 1% wt. Similarly, the resulting carbon fibres also showed a maximum improvement at a concentration of 1% wt. The decrease in strength seen at levels beyond 1% wt. was attributed to a lack of preferred orientation of the clay platelets along the fibre axis, increased fibre diameter, and the presence of micro-voids in the carbon fibres. It was suggested that the diameter reliance of oxidative thermostabilization for lignin is low and the addition of nondiffusive nanomaterials at higher levels to lignin could reduce the rate of oxidative thermostabilization of lignin fibres because they provide a substantial barrier to air diffusion.

The possibility of using carbon nanotubes for increasing the strength, modulus, thermal conductivity and electrical conductivity of different lignins used to produce carbon fibre was examined by Baker et al. (2011). Study with an organosolv lignin, a purified hardwood kraft lignin, and a blend containing a softwood kraft lignin and the hardwood lignin (purified), showed that the carbon nanotubes could be added at a higher level of 15 wt% before fibres could not be melt-spun. In many cases at an optimal level of CNTs, the nanotubes increased the fibre spinning process by the increased heat capacity of the composite, that allowed the fibre to remain molten a greater distance on the spin-line, thus allowing increased stretching and more fine diameter fibres to be spun. The lignin composite fibres were much stronger than fibres produced using lignin alone. But, after oxidative thermostabilization and subsequent carbonization, the carbon fibres showed 50% increase in modulus and only a 20% increase in tensile strength as compared to the neat lignin-based carbon fibres. The reason was ascribed to low interfacial adhesion between the nonfunctionalized CNTs and the lignin carbons.

Wohlmann et al. (2010) reported methods by which lignins were derivatized through their free hydroxyl groups to give plasticizing derivatives attached via ester, ether and/or urethane functions.

A method of producing activated carbon fibres using phenol-formaldehyde resin technology was reported by Shen et al. (2011).

Maradur et al. (2012) prepared a copolymer using a hardwood lignin and acrylonitrile with acrylonitrile: lignin ratio from 5:5 to 8:2. The lignin- polyacrylonitrile copolymer was dissolved in DMSO and converted into fibres through wet-spinning using a coagulation bath containing water. Copolymerization was achieved using a two-step radical polymerization process; in the first step, the oligomerization of acrylonitrile took place; the second step involved the addition of peroxide activated lignin to the acrylonitrile oligomers. The product, containing up to 50 wt% lignin, could be spun from a 16% copolymer DMSO solution into fine fibres of around 15 mm diameter. Oxidative thermostabilization was conducted isothermally at 280 °C, and carbonization proceeded at 5 °C/min to 800 °C without fibre fusion, giving carbon fibres of around 10 mm diameter. Details concerning the synthesis, some characteristics of the copolymer, fibre spinning and conversion were described. Further work to make improvements to the process and characterize the resulting carbon fibres is in progress (Yang et al. 2012).

The possibility of using lignocellulosic and liquid wood precursors have been examined (Lehmann et al. 2012a, b; Uraki and Kubo 2006). Although these are not strictly lignin-based carbon fibres, they represent an interesting class of material, as they could provide a way to low-cost carbon fibre through the use of low-value lignocellulosic streams, potential biorefinery waste streams and/or sawdust. Varying levels of lignin additive were added to a polymer solution of a cellulose/ cellulose derivative that is solution spun into a coagulation bath to form spooled continuous fibres. The solution forming solvents were ionic liquids— N-methylmorpholine-N-oxide, and other solvents common to cellulose and cellulose derivative dissolution.

A process was suggested by Ma and Zhao (2011) in which powdered wood is first mixed in phenol to which phosphoric acid has been added at 160 °C for 2.5 h. The liquid wood is then polymerized by the addition of hexamethylenetetramine at a level of 5 wt% and heated to cause cross linking. The solution is then spun and fibre stabilization is done by soaking the fibres in a solution containing formaldehyde and hydrochloric acid at 90 °C for 2 h. The tensile strengths and moduli approached 1.7 GPa and 176 GPa, respectively (Ma and Zhao 2010; Liu and Zhao 2012). Similar syntheses. have been reported by Lin et al. (2013a) and Kato et al. (2012) but it is not clear if these processes could be economically or environmentally attractive.

Carbon nanofibres with of small diameters of 200 nm were successfully prepared from lignin using electrospinning (Lallave et al. 2007). These researchers formulated a method by which an Alcell (organosolv lignin) dissolved in ethanol at ratios of 1:1 w/w could be electrospun coaxially to give lignin fibres in the range of 400 nm to 2 mm diameter. The yield of carbon nanofibres was 31.6% (based on electrospun lignin fibre), and had surface areas of up to 1200 $m^2$/g.

Ruiz-Rosas et al. (2010) manufactured lignin fibres in a single-step procedure utilizing electrospinning. They reported the formation of both lignin-based and platinum-containing carbon nanofibres while measuring the loss of small molecules during thermal conversion treatments for providing some mechanistic insights. Elemental composition, adsorption/desorption isotherms and functional groups, were characterized as a function of conversion ordinate and platinum contents.

A method was reported by Hosseinaei and Baker (2012a, b). They described the purification and preparation of a commercial lignin by solvent extraction to give a high Tg lignin that can be electrospun, rapidly oxidatively thermostabilized and carbonized. These researchers produced nanofibre mats with smooth surfaces and without defects from a softwood kraft lignin purified by solvent extraction before electrospinning. The structural alignment of the lignin-based carbon fibre mats was found to be very important in further improving the mechanical strength of the material (Lin et al. 2013b).

Electrospinning of PAN/lignin composite submicron fibres was reported by Seo et al. (2011). Solutions containing between 100:0 and 80:20 w/w PAN/lignin were electrospun under exactly similar conditions to produce fibrous mats of varying morphology. Eelectron beam irradiation was used for oxidative thermostabilization and carbonization was effected by thermal treatment to 1000 °C at a rate of 10 °C/min, to give fairly uniform fibres of 300 nm average diameter. The carbon fibre mats showing best performance were produced using a 50:50 PAN/ lignin solution, 15 kV, and with a target distance of 100 mm for producing the fibre mat. Radiation dose of 2000 kGy was used to render the fibres infusible prior to carbonization.

Chatterjee et al. (2014) reported a process for the conversion of solvent-extracted Alcell hardwood and Kraft softwood lignins into carbon fibres with a morphology optimized for advanced energy storage applications, such as anode materials for lithium or sodium ion batteries. Lignin carbon fibres were produced from unmodified and acid anhydride modified Alcell hardwood and softwood lignin precursors. These materials were melt-processed into lignin fibres and optimal extrusion parameters were determined for each type of lignin precursor. Although melt processing of softwood lignins is generally considered not feasible, modification of Kraft softwood lignin with phthalic or acetic anhydride resulted in a precursor which was found to be melt extruded and spun into fibres. Analysis of the precursors showed clear differences in the chemical composition of hardwood and softwood lignin and as a result of chemical modifications. Chemical modification of Kraft softwood lignin results in a mass distribution similar to Alcell hardwood lignin. All lignin fibres were oxidatively stabilized and carbonized under identical conditions.

The summary of properties of carbon fibre produced from different types of lignin are presented in Table 7.4.

**Table 7.4** Properties of carbon fibre from different types of lignin

| Precursor type | Diameter (μm) | Elongation (%) | Modulus (Gpa) | Tensile strength (Mpa) |
|---|---|---|---|---|
| Types of lignin | N/A | N/A | N/A | 150–800 |
| Steam exploded hardwood | 8 ± 3 | 1.6 ± 0.2 | 40.7 ± 6.3 | 660 ± 230 |
| Steam exploded hardwood | | 1.2 | | 450 |
| Organosolv hardwood | 14 ± 1 | 1.0 ± 0.3 | 39.1 ± 13.3 | 355 ± 53 |
| Organosolv softwood | 84 ± 15 | 0.7 ± 0.1 | 3.6 ± 0.4 | 26.4 ± 3.1 |
| Kraft hardwood | 46 ± 8 | 1.1 ± 0.2 | 40 ± 11 | 422 ± 80 |
| Oganic purified hardwood | 10 ± 1 | | 28.6 ± 3.2 | 520 ± 182 |
| Oganic purified hardwood | | 2.0 | 82.7 | 1070 |
| Softwood and hardwood kraft | 36–78 | 0.78–1.20 | 25–33 | 233–377 |
| lignosulfonate- AN copolymer | 12–20 | | 540 | |
| PAN | 5–10 | 2 | 100–500 | 3000–7000 |
| Mesophase pitch | 5–15 | 0.6 | 200–800 | 1000–3000 |

Otani et al. (1969), Sudo and Shimizu (1992), Sudo et al. (1993), Uraki et al. (1995), Kadla et al. (2002a, b), Baker (2011), Xia et al. (2016), Luo (2004), Kubo et al. (1998), Zhang (2016)

# References

Attwenger A (2014) Value-added lignin based carbon fibre from organosolv fractionation of poplar and switchgrass. Thesis, Master of Science, University of Tennessee—Knoxville, USA

Bajpai P (2013) Update on carbon fiber. Smithers Rapra, UK

Baker FS (2010a) Low cost production of carbon fibre from sustainable resource materials: utilization for structural and energy efficiency applications. Presentation to Ontario Bio-Auto Council, April 8–9, 2010. SEM images with higher definition courtesy of Paul Menchhofer of Oak Ridge National Laboratory

Baker FS (2010b) Presentation at 2010 DOE hydrogen program and vehicle technologies annual merit review and peer evaluation meeting, June 7–11, 2010. Available at: http://www1.eere. energy.gov/vehiclesandfuels/pdfs/merit_review_2010/lightweight_materials/lm005_baker_2010_o.pdf

Baker FS (2011) Utilization of sustainable resources for materials for production of carbon fibre structural and energy efficiency applications, nordic wood biorefinery conference, March 22–24, 2011, Stockholm Sweden

Baker DA, Rials TG (2013) Recent advances in low-cost carbon fibre manufacture from lignin. J Appl Polym Sci 130:713–28

Baker DA, Gallego NC, Baker FS (1969) On the characterization and spinning of an organic-purified lignin toward the manufacture of low-cost carbon fibre. J Appl Polym Sci 2012 124 (1):227

Baker FS, Griffith WL, Compere AL (2005) Low-cost carbon fibres from renewable resources. Automotive light weight materials, FY 2005 Process report 187–196

Baker DA, Gallego NC, Baker, FS (2008a) Extended abstract. In: Book of abstracts of the fibre society 2008 fall conference, Industrial Materials Institute, Montreal, Canada, October 1–3, 2008. Available at: http://www.thefibresociety.org/Assets/Past_Meetings/PastMtgs_Home.html

Baker FS, Gallego NC, Baker DA (2008b) Progress report for lightweighting materials, Part 7.A. 2008. Available at: https://www1.eere.energy.gov/vehiclesandfuels/pdfs/lm_08/7_low-cost_carbon_fibre.pdf

Baker DA, Baker FS, Gallego, NC (2009a) Extended abstract. In: Book of abstracts of the fibre society 2009 fall conference, University of Georgia, Athens, GA, USA, October 27–30, 2009. Available at: http://www.thefibresociety.org/Assets/Past_Meetings/PastMtgs_Home.html

Baker FS, Gallego NC, Baker DA (2009b) Progress report for Lightweighting Materials, Part 7.A. 2009. Available at: http://www1.eere.energy.gov/vehiclesandfuels/pdfs/lm_09/7_low-cost_carbon_fibre.pdf

Baker FS, Baker DA, Gallego NC (2010a) Carbon fibre from lignin. In: Proceeding Carbon 2010, Clemson, SC, July 11–16

Baker FS, Gallego NC, Baker DA (2010b) Progress report for Lightweighting Materials, Part 3. 32–34, 2010. Available at: http://www1.eere.energy.gov/vehiclesandfuels/pdfs/lm_10/2010_lightweighting_materials.pdf

Baker DA, Gallego NC, Baker FS (2010c) SAMPE'10 Conference and exhibition, Seatle, WA, May 17–20 2010

Baker FS, Baker DA, Menchhofer PA (2011) Carbon nanotube enhanced precursor for carbon fibre production and method of making a CNT-enhanced continuous lignin fibre. US Patent Application 2011285049, 2011, to be assigned to Oak Ridge National Laboratory

Baker DA, Harper DP, Rials TG (2012) Carbon fibre from extracted commercial softwood lignin. In: Book of abstracts of the fibre society 2012 fall conference, Boston, The Fibre Society, pp 17–18

Berlin A (2011) Carbon fibre compositions comprising lignin derivatives. WO 2011097721 A1, 2011 assigned to Lignol Innovations Ltd

Bissett PJ, Herriott CW (2012) Lignin/polyacrylonitrile-containing dopes, fibres and production methods. US 20120003471 & WO 2012003070, 2012, assigned to Weyerhaeuser NR Company

Braun JL, Holtman KM, Kadla JF (2005) Lignin-based carbon fibres: oxidative thermostabilization of kraft lignin. Carbon 43:385–394

Brodin I, Ernstsson M, Gellerstedt G, Sjöholm E (2012) Oxidative stabilization of Kraft lignin for carbon fibre … Holzforschung 66:141–7

Chatterjee S, Jones EB, Clingenpeel AC, McKenna AM, Rios O, McNutt NW, Keffer DJ, Johs A (2014) Conversion of lignin precursors to carbon fibres with nanoscale graphitic domains. ACS Sustainable Chem Eng 2014(2):2002–2010

Dong X, Lu C, Zhou P, Zhang S, Wang L, Li D (2015) Polyacrylonitrile/lignin sulfonate blend fibre for low-cost carbon fibre. RSC Adv 2015(5):42259–42265

Donnet JB, Bansal RC (1990) Carbon fibres, 2nd edn. Marcel Dekker, New York, USA, pp 1–145

Eckert RC, Abdullah Z (2008) Carbon fibres from kraft softwood lignin. Application US 20080317661, assigned to Weyerhaeuser Company

Frank E, Hermanutz F, Buchmeiser MR (2012) Carbon fibres: precursors, manufacturing, and properties. Macromol Mater Eng 297:493–501

Frank E, Steudle LM, Ingildeev D, Spörl JM, Buchmeiser MR (2014) Carbon fibres: precursor systems, processing, structure, and properties. Angew Chem Int Ed 53:2–39

Fukuoka Y (1969) Carbon fibre made form lignin. Jpn Chem Q 5(3):63–66

Gellerstedt G, Sjoholm E, Brodin I (2010) The wood-based biorefinery: a source of carbon fibre? Open Ag J 3:119

Gould AM (1974) Manufacture of carbon fibre GB Pat 1,358,164

Hosseinaei O, Baker DA (2012a) Electrospun carbon nanofibres from kraft lignin. http://www. thefibresociety.org/httpdocs/Assets/Past_Meetings/BooksOfAbstracts/2012_Fall_Abstracts.pdf

Hosseinaei O, Baker DA (2012b) Extended abstract in Book of Abstracts of The Fibre Society 2012 Fall Conference, Boston Convention and Exhibition Center, Boston, USA, November 7–9, 2012. Available at: http://www.thefibresociety.org/Assets/Past_Meetings/PastMtgs_Home. htm

Ichikawa H, Yokoyama A, Nankjima N (1992) JP Pat 4,194,029

Ito K, Shigemoto T (1989) JP Pat 1,239,114

Kadla JF, Kubo S (2005) Lignin-based polymer blends: analysis of intermolecular interactions in lignin-synthetic polymer blends 98:1437–44

Kadla JF, Kubo S, Gilbert RD, Venditti RA, Compere AL, Griffith WL (2002) Lignin-based carbon fibres for composite fibre applications. Carbon 40(15):2913–2920

Kato O, Ito K, Matsunaga K, Sakanishi K (2012) Manufacture of carbon fibres from solid woody materials. JP 2012117162, 2012, assigned to AIST and Tokai Carbon Co. Ltd

Kim UC, Jung JC, Jin SU (2011) Method for manufacturing lignin melt with high processibility. KR 2011108944, 2011, assigned to Kolon Industries

Kubo S, Kadla JF (2004) Poly (ethylene oxide)/organosolv lignin blends: relationship between thermal properties, chemical structure, and blend behavior. Macromolecules 37:6904–11

Kubo S, Kadla JF (2005) Lignin-based carbon fibres: effect of synthetic polymer blending on fibre134 properties. J Polym Environ 13:97–105

Kubo S, Kadla JF (2005a) Lignin-based carbon fibres: effect of synthetic polbiocomposite materials, natural fibres, biopolymers, and biocomposites, pp 671–697

Kubo S, Kadla JF (2005b) Kraft/lignin/poly(ethylene oxide) blends: effect of lignin structure on miscibility and hydrogen bonding. J Appl Polym Sci 98:1437–1444

Kubo S, Ishikawa M, Uraki Y, Sano Y (1997) Preparation of lignin fibres from softwood acetic acid lignin relationship between fusibility and the chemical structure of lignin. Mokuzai Gakkaishi 43:655–662

Kubo S, Uraki Y, Sano Y (1998) Preparation of carbon fibres from softwood lignin by atmospheric acetic acid pulping. Carbon 36(7–8):1119–1124

Kadla JF, Kubo, S, Gilbert, RD, Venditti, RA (2002b) Lignin-based carbon fibres, chemical modification, properties, and usage of lignin.In: Hu TQ (ed) Kluwer Academic/Plenum, New York, pp 121–137

Kurimoto Y, Doi S, Tamura Y (1999) Species effects on wood-liquefaction in polyhydric alcohols. Holzforschung 53:617–622

Lallave M, Bedia J, Ruiz-Rosas R, Rodriguez-Mirasol J, Cordero T, Otero JC (2007) Filled and hollow carbon nanofibres by coaxial electrospinning of Alcell lignin without binder polymers. Adv Mater 2007(19):4292–6

Lehmann A, De Ebeling H, Fink HP (2012a) Application WO 2012/156443, 2012. Method for the production of lignin-containing precursor fibres and also carbon fibres WO 2012156443 A1

Lehmann A, Ebeling H, Fink HP (2012b) Method for the production of lignin-containing precursor fibres and also carbon fibres. International patent application, WO 2012/ 156441, to be assigned to Fraunhofer Institute

Liu W, Zhao G (2012) Effect of temperature and time on microstructure and surface functional groups of activated carbon fibres prepared from liquefied woods. Bioresources 7:5552

Lin J, Kubo S, Yamada T, Koda K, Uraki Y (2012) Chemical thermostabilization for the preparation of carbon fibres from softwood lignin. Bioresources 7:5634–46

Lin J, Shang JB, Zhao G (2013a) Preparation and characterization of liquefied wood based primary fibres. Carbohydr Polym 91:224

Lin L, Li Y, Ko FK (2013b) Fabrication and properties of lignin based carbon nanofibre. J Fibre Bioeng Inform 6:335–47

Luo J (2004) Lignin based carbon fibre. Thesis Master of Science, Chemical Engineering University of Maine. USA, 2004

Luo J, Genco J, Cole B, Fort R (2011) Lignin recovered from the near-neutral hemicellulose extraction processas a precursor for carbon fibre. Bioresources 6:4566–4593

Ma X, Zhao G (2010) Preparation of carbon fibres from liquefied wood. Wood Sci Technol 2010 (44):3–11

Ma X, Zhao GJ (2011) Variation in the microstructure of carbon fibre s prepared from liquefied wood during carbonization Appl. Polym Sci 121:3525

Mansmann M (1974) Stable lignin fibres. GB Pat 1,359,764

Maradur SP, Kim CH, Kim SY, Kim B, Kim WC, Yang KS (2012) Preparation of carbon fibres from a lignin copolymer with polyacrylonitrile. Synth Met 162:453–459

Minus ML, Kumar S (2005) The processing, properties and structure of carbon fibres. JOM 57 (2):52–58

Norberg I (2012) Carbon fibres from kraft lignin (Doctoral Thesis) TRITA-CHE Report 2012:13, Stockholm, Sweden, Royal Institute of Technolog, 2012, p 97 (ISBN 9789175012834)

Norberg I, Nordström Y, Drougge R, Gellerstedt G, Sjöholm E (2012) A new method for stabilizing softwood kraft lignin fibres for carbon fibre production. J Appl Polym Sci 128 (6):3824–3830

Nordström Y (2012) Development of softwood kraft lignin based carbon fibres. Licentiate Thesis, Department of Engineering Sciences and Mathematics Luleå University of Technology, 2012

Nordstrom Y, Norberg I, Sjoholm E, Drougge R (2012) J Appl Polym Sci 2012. doi:10.1002/APP. 38795. A new softening agent for melt spinning of softwood kraft lignin

Nordström Y, Norberg I, Sjöholm E, Drougge R (2013a) A new softening agent for melt spinning of softwood kraft lignin. J Appl Polym Sci 129:1274–1279. 132

Nordström Y Joffe, Sjöholm RE (2013b) Mechanical characterization and application of Weibull statistics to the strength of softwood lignin-based carbon fibres. J Appl Polym Sci 130:3689–3697

Norgren M, Edlund H (2014) Lignin: recent advances and emerging applications. Curr Opin Colloid and Interface Sci 19:409–416

Otani S (1967) Tanso, 50, p 32

Otani S. Fukuoka Y, Igarashi B, Sasaki K (1966) FR Pat 1,458,725

Otani S, Fukuoka Y, Igarashi B, Sakaki K (1969) Method for producing carbonized lignin fibre (Nippon Kayaku Kk), US Pat. 3461082

Peebles LH (1994) Carbon fibres: formation, structure, and properties, 1st edn. CRC Press, Boca Raton, FL, USA, p 3

Prauchner MJ, Pasa VMD, Otani S, Otani C (2005) Biopitch-based general purpose carbon fibres: processing and properties. Carbon 43(3):591–597

Qiao WM, Huda M, Song Y, Yoon SH, Korai Y, Mochida I (2005) Carbon fibres and films based on biomass resins. Energy Fuels 19(6):2576–2582

Qin W, Kadla J (2011) Effect of organoclay reinforcement on lignin-based carbon fibres. Ind Eng Chem Res 50:12548–12555

Qin W, Kadla JF (2012) Carbon fibres based on pyrolytic lignin. J Appl Polym Sci 126:E203–E212

Ruiz-Rosas R, Bedia J, Lalleve M, Loscertales IG, Barrero A, Rodríguez-Mirasol J (2010) The production of submicron diameter carbon fibres by the electrospinning of lignin. Carbon 48:696–705

Sazanov YN, Fedorova GN, Kulikova EM, Kostycheva DM, Novoselova AV, Gribanov AV (2007) Cocarbonization of polyacrylonitrile with lignin. Russ J Appl Chem 80:619. doi:10. 1134/S1070427207040209

Sazanov YN, Kostycheva DM, Fedorova GN, Ugolkov VL, Kulikova EM, Gribanov AV (2008) Composites of lignin and polyacrylonitrile as carbon precursors. Russ J Appl Chem 81 (7):1220–1223

Schmidt JA, Rye CS, Gurnagul N (1995) Lignin inhibits auloxidative degradation of paper. Polym Deg Stab 49:291–297

Scholze B, Hanser C, Meier D (2001) Characterization of the water-insoluble fraction from fast pyrolysis liquids (pyrolytic lignin). J Anal Appl Pyrolysis 60:41–54

Seo DK, Jeun JP, Kim HB, Kang PH (2011) Preparation and characterization of the carbon nanofibre mat produced from electrospun pan/lignin precursors by electron beam irradiation. Rev Adv Mater Sci 2011(28):31–34

Sevastyanova O, Qin W, Kadla JF (2010) Effect of nanofillers as reinforcement agents for lignin composite fibres. J Appl Polym Sci 117(5):2877–2881

Seydibeyoglu MO (2012) A novel partially biobased PAN-lignin blend as a potential carbon fibre precursor. J Biomed Biotechnol 2012:598324. doi:10.1155/2012/598324

Shen Q, Zhang T, Xu Y (2007) Carbon nanofibres carbonized from lignin/polymer fibres and preparation thereof. CN 101078137

Shen Q, Zhang T, Zhang WX, Chen S, Mezgebe MJ (2011) Lignin-based activated carbon fibres and controllable pore size and properties. Appl Polym Sci 121:989–994

Sjoholm E, Gellerstedt G, Drougge R, Brodin I (2012) Method for producing a lignin fibre. Application WO 2012/112108 A1, 2012, to be assigned to Innventia AB

Sudo K, Shimizu K (1987) JP Pat 62(110):922

Sudo K, Shimizu K (1989) JP Pat 0,136,618

Sudo K, Shimizu K (1992) A new carbon fibre from lignin. J Appl Polym Sci 44(1):127–134

Sudo K, Shimizu K (1994). Method for manufacturing lignin for carbon fibre spinning. U.S Pat 5,344,921

Sudo K, Okoshi, Shimizu, K (1988) Am Chem Soc Abstracts V195, 107-Cell 1988

Sudo K, Shimizu K, Nakashima N, Yokoyama A (1993) A new modification method of exploded lignin for the preparation of a carbon fibre precursor. J Appl Polym Sci 48(8):1485–1491

Tagawa T, Miyata T (1997) Size effect on tensile strength of carbon fibers. Mater Sci Eng A 238:336–342

Takanori M, Shinya K, Eiichi Y, Otani A (2010) Manufacture of long, ultrafine, and unbranched carbon fibres from lignin derivatives and using fibreization aids. JP 2010242248, 2010, assigned to Teijin Ltd

Tomani P (2010) The lignoboost process. Cellul Chem Technol 44:53

Uraki Y, Kubo S (2006) Mokuzai Gakkaishi 52:337

Uraki Y, Kubo S, Nigo N, Sano Y, Sasaya T (1995) Preparation of carbon fibres from organosolv lignin obtained by aqueous acetic acid pulping. Holzforschung 49:343–350

Uraki Y, Kubo S, Kurakami H, Sano Y (1997) Activated carbon fibres from acetic acid lignin. Holzforschung 51:188–192

Uraki Y, Nakatani A, Kubo S, Sano YJ (2001) Preparation of activated carbon fibre with large specific area from softwood acetic acid lignin. Wood Sci 47:465

Wohlmann B, Woelki M, Ebert A, Engelmann G, Fink HP (2010) Lignin derivative, shaped body comprising the derivative, and carbon fibres produced from the shaped body. WO 2010081775, assigned to Toho Tenax Europe GmbH and Fraunhofer Institute of Germany

Wohlmann B, Woelki M, Stuesgen S (2012) Thermoplastic lignin for producing carbon fibres. WO 2012038259, 2012, assigned to Toho Tenax Europe GMBH

Xia K, Ouyang Q, Chen Y, Wang X, Qian X, Wang L (2016) Preparation and characterization of lignosulfonate-acrylonitrile copolymer as a novel carbon fibre precursor. ACS Sustain Chem Eng 2016(4):159–168

Yang GS, Yoon, JH, Nillesh SL, Kim YC (2011) Lignin based complex precursor for carbon fibres and method for preparing lignin-based carbon fibres using it. KR 2011116604, 2011, assigned to Industry Foundation of Chonnam National University

Yang GS, Mahradu S, Kim, YC (2012) Method for manufacturing carbon fibres using lignin based copolymer, and carbon fibres using the same. KR 2012109227, assigned to Industry Foundation of Chonnam National University, South Korea

Zhang M (2016) Carbon Fibres Derived from Dry-Spinning of Modified Lignin Precursors. Ph. D thesis, Chemical engineering, Clemson University

# Chapter 8
# Conversion of Lignin to Carbon Fibre

**Abstract** Methods for conversion of lignin to carbon fibre are presented in this chapter. Direct melt spinning is the most common method for lignin fibre. Electrospinning is another important method and is an alternative to the melt-spinning technique.

**Keywords** Lignin · Carbon fibre · Direct melt spinning · Electrospinning · Conversion of lignin

There are three main methods for fibre formation from polymers. They are as follows:

- Wet Spinning
- Dry Spinning
- Melt Spinning.

## 8.1 Wet Spinning

Wet spinning is used for fibre-forming substances that have been dissolved in a solvent and is the oldest process. The spinnerets are submerged in a chemical bath and as the filaments come out they precipitate from solution and solidify. As the solution is extruded directly into the precipitating liquid, this process is called wet spinning.

## 8.2 Dry Spinning

In case of dry spinning, instead of precipitating the polymer by dilution or chemical reaction, solidification is obtained by evaporating the solvent in a stream of air or inert gas. The filaments do not come in contact with a precipitating liquid. Therefore, the need for drying is eliminated and the recovery of solvent is easier.

© The Author(s) 2017                                                                                               63
P. Bajpai, *Carbon Fibre from Lignin*, SpringerBriefs in Materials,
DOI 10.1007/978-981-10-4229-4_8

## 8.3   Melt Spinning

In this case, the fibre-forming substance is melted for extrusion through the spinneret and then directly solidified by cooling. Melt spinning is the most rapid, convenient and commonly used method of forming polymeric fibres. With melt-spinning techniques the use of solvents can be significantly reduced.

All these methods involve pumping the melt or solution of the polymer through holes in a spinneret. The spinneret holes match the desired filament count of the carbon fibre (Chung 1994; McConnell 2008). Melt spinning is not possible in case of PAN because PAN decomposes below its melting temperature. PAN-based fibres are mostly produced by either wet spinning or melt-assisted spinning process. In both the processes organic solvent is used. In case of melt-assisted spinning, a solvent in the form of hydrating agent to decrease the melting point of PAN is used. The polymer can be melted with a low melting point and pumped through a spinneret (Chung 1994). In case of wet spinning, the dissolved precursor is immersed in a liquid coagulation bath and extruded through holes in a spinneret. The wet-spun fibre is drawn by using rollers through a wash for removing the excess coagulant. The fibres are then dried and stretched to the correct fibre specification (McConnell 2008).

For the lignin fibre, the direct melt spinning is the most common method. Its production process requires the extrusion of only the pure polymer precursor directly into fibre form, removing the requirement of extra expense of solvent recovery and providing a more environmentally sound solution. Lignin should reach purity of greater than 99% for melt spinning and the particles size no larger than 1 micron (Eberle 2012). Lignin should be prepared in such a way that allows low melt flow temperature so that it can be melt-spun without polymerizing during extrusion. It also requires a high glass transition temperature for the fibre to stabilize at an acceptable rate. The glass transition temperature is maintained above the oxidation temperature so that the fibre crosslinks and stabilizes without infusion. Following conditions should be carefully controlled for obtaining carbon fibre with required strength (Fig. 8.1).

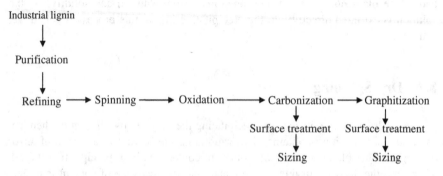

**Fig. 8.1** Production steps involved in the production of carbon fibre from lignin. Based on Baker and Rials (2013), Chen (2014)

- Spinning conditions
- Treatment temperatures
- Temperature ramping profiles.

Electrospinning is another important carbon fibre production method and is an alternative to the melt spinning technique. It is used to produce fibres in the diameter from 0.03 to 1 mm range and is an effective technique for the fabrication of fibres in submicron diameter range, 0.1 μm, from polymer solutions or melts. Typically, an electric potential is used between a droplet of polymer solution at the end of a capillary and a grounded collector. When the applied electric field overcomes the surface tension of the droplet, a charged jet of polymer solution or melt is ejected. The jet grows longer and thinner until it solidifies or collects on the collector. This fibre morphology is controlled by the following properties (Lin et al. 2013):

- Solution conductivity
- Concentration
- Viscosity
- Molecular weight
- Applied voltage.

Lignin is a thermoplastic polymer making melt spinning an applicable method to spin fibres. In contrast to kraft lignin, organosolv lignin contains only a very small amount of inorganic material, which provides good melt-spinning opportunities. Inorganic materials are contaminants that will degrade carbon fibre properties and are a result of the lignin recovery process. Melt spinning of lignin was mentioned first by Otani et al. (1969) describing several methods of forming fibre from lignin using a one-pot melt-spinning method. Since that time, many achievements arose and current technical improvements enable faster and relatively easy handling (Attwenger 2014). Baker et al. (2012) produced melt-spun lignin fibre from kraft hardwood lignin and also from an organic purified hardwood lignin. Organic solvent purification process was used for dissolving the lignin from most impurities to improve its melt spinnability. The fibres were then stabilized and carbonized to obtain lignin-based carbon fibre. The major disadvantage was the slow heating rates for fibre stabilization. The oxidative stabilization was achieved at heating rates smaller than 0.05 °C/min. Decreasing heating rates substantially increased the glass transition temperature and also allowed the lignin to crosslink and deliver stabilized fibres. The mechanical properties of the carbon fibres were poor, and, therefore, research is required to decrease stabilization time and improving their properties.

Zhang (2016) performed dry spinning over a range of concentrations (1.85–2.15 g/ml acetone) and appropriate temperatures (25–50 °C). All of the resulting dry-spun lignin fibres showed a crenulated surface pattern, with increased crenulation obtained for fibres spun at higher temperatures. Presence of some doubly convex and sharp crevices was observed on fibres produced from solutions containing reduced concentrations (1.85 and 2.00 g lignin/mL solvent). Contrary to this, no crevices were seen on the fibres obtained from the concentrated solution

(2.15 g/mL), possibly because of the reduced extent of solvent out-diffusion. Dry spinning at room temperature was conducted to obtain fibres with relatively smooth surface, but the pressure drop was very high. These results clearly establish temperature/concentration combinations for dry spinning of Ace-SKL. About 30% higher surface area was obtained in the crenulated lignin fibres (as compared with equivalent circular fibres), showing the potential benefits of such biomass-derived fibres in providing larger fibre/matrix bonding area when used in composites.

After lignin is melt-spun into fibre under an inert atmosphere, it is then oxidatively thermostabilized and carbonized. The integrity of the lignin fibre during this process depends on its ability to crosslink, so that the glass transition of the material is maintained above the process temperature, finally rendering it infusible. The process is complex and careful control of the following conditions is required for obtaining the carbon fibre of superior strength:

- Lignin spinning conditions
- Treatment temperatures
- Ramping profiles.

Lignin, which is naturally partially oxidized, requires critical control of the melt-spinning step. The lignin should be produced in such a way to have a low enough melt flow temperature for it to be melt-spun without polymerizing during extrusion, but a high enough glass transition for fibre stabilization to proceed at an acceptable rate. Thus opportunity exists for producing carbon fibre from lignin by melt extrusion. But lignin has advantages over other precursors like MPP and PAN:

- Very inexpensive
- A renewable product
- Already substantially oxidized so that it can be oxidatively thermostabilized at potentially much higher rates than either MPP or PAN.

# References

Attwenger A (2014) Value-added lignin-based carbon fibre from organosolv fractionation of poplar and switchgrass. A thesis presented for the Master of Science Degree, The University of Tennessee, Knoxville

Baker DA, Gallego NC, Baker FS (2012). On the characterization and spinning of an organic-purified lignin toward manufacture of low-cost carbon fiber. J Appl Polym Sci 227–234

Baker DA, Rials TG (2013) Recent advances in low-cost carbon fibre manufacture from lignin. J Appl Polym Sci 130:713–28

Chen MCW (2014) Commercial viability analysis of lignin based carbon fibre. Master thesis, Simon Fraser University

Chung DDL (1994) Carbon fibre composites. Butterworth-Heinemann

Eberle C (2012, April 12) Carbon fibre from lignin. Oak Ridge, TN. Retrieved from http://www.cfcomposites.org/PDF/Breakout_Cliff.pdf

Lin L, Yingjie L, Ko FK (2013) Fabrication and properties of lignin based carbon nanofibre. J Fibre Bioeng Inform 6(4). doi:10.3993/jfbi12201301

McConnell V. (2008, December 19). The making of carbon fibre. CompositesWorld. Retrieved 10 May 2014 from http://www.compositesworld.com/articles/the-making-of-carbonfibre

Otani S, Fukuoka Y, Igarashi B, Sasaki K (1969, August 12) Method for producing carbonized lignin fibre

Zhang M (2016) Carbon fibres derived from dry-spinning of modified lignin precursors. Ph. D thesis, Chemical engineering, Clemson University

# Chapter 9
# Future Directions

**Abstract** Future directions of carbon fibre industry are presented in this chapter.

**Keywords** Lignin-based carbon fibre · Lignin · Carbon fibre · Wood-based biorefinery · Low-cost alternative · Ppetroleum-based precursors

Although there has been a continuous growth of carbon fibre production over the last few years, large-volume applications of carbon fibres, have been impeded by the high fibre costs and the lack of techniques for the high-speed production of composites (Chung 1994; Donnet and Bansal 1990; Minus and Kumar 2005, 2007; Watt 1985; Baker and Rials 2013; Bajpai 2013; Frank 2014). However, present investigations and future developments might well change the market and establish carbon fibres as a mass product similar to other synthetic fibres or even metals. Renewable raw materials are particularly interesting sources of carbon fibres.

Lignin-based carbon fibre is the most value-added product from a wood-based biorefinery and has the potential of providing a low-cost alternative to petroleum-based precursors to manufacture carbon fibre, which can be combined with a binding matrix for producing a structural material with much greater specific strength and stiffness compared to conventional materials such as aluminium and steel. The market for carbon fibre is projected to grow exponentially to fill the requirements of clean energy technologies such as wind turbines and to improve the fuel economies in vehicles through lightweighting. In addition to cellulosic biofuel production, lignin-based carbon fibre production coupled with biorefineries may provide \$2400–\$3600 added value dry $Mg^{-1}$ of biomass for vehicle applications (Langholtz et al. 2014). Compared to production of ethanol alone, the addition of lignin-derived carbon fibre could increase biorefinery gross revenue by 30–300%. Using lignin-derived carbon fibre in 15 million vehicles per year in the United States could reduce fossil fuel consumption by 2–5 billion litres $year^{-1}$, reduce carbon dioxide emissions by about 6.7 million Mg $year^{-1}$, and realize fuel savings through vehicle lightweighting of \$700–\$1600 per Mg biomass processed. The value of fuel savings from vehicle lightweighting becomes economical at carbon

P. Bajpai, *Carbon Fibre from Lignin*, SpringerBriefs in Materials,
DOI 10.1007/978-981-10-4229-4_9

fibre price of $6.60 \text{ kg}^{-1}$ under current fuel prices, or $13.20 \text{ kg}^{-1}$ under fuel prices of about $1.16 \text{ l}^{-1}$.

There are many problems in providing lignins suitable for manufacture of carbon fibre. No demonstration has been made yet of suitable lignins being processed into carbon fibre which satisfy strength requirements and also the cost objectives. This is because of the unavailability of lignins with suitable properties so that they can be melt spun into fibre; converted rapidly to carbon fibre at low cost. This requires a low Tg lignin to be melt spun, requires a high Tg lignin to assure low cost, and requires minimal processing cost of the lignin before fibre spinning; for a melt-spinning process, and may therefore be high.

Using alternative fibre spinning methods which are not bound by the rigorous technical requirements required for the traditional melt spinning of fibres, cost reductions could be obtained. However, the provision of refined or improved lignins may yet provide the necessary lignin performance qualities required for traditional melt spinning, or indeed the use of any fibre forming method to produce higher strength carbon fibre (Baker et al. 2012; Morck et al. 1986). In consideration of the methods used to manufacture lignin, lignin products will all require purification and refining to be suitable for use in manufacture of carbon fibre. This has been found to be true for even an advanced organosolv process which provides high purity, well defined lignins directly and with much more appropriate Tg and Ts properties than competing organosolv technologies (Black et al. 1998; Bozell et al. 2011). In the context of lignocellulosic biorefining towards production of advanced fuels, the use of organosolv processes has been found to be cost competitive with pretreatment and other fractionation technologies and with the added provision of relatively pure lignins (Bozell 2010a, b; Michels and Wagemann 2010; Black et al. 1998; Bozell et al. 2011; Lignol Energy Corporation 2010, 2011). Nevertheless, in order to direct particular pretreatment, fractionation, or lignin refining techniques for obtaining specific lignins for manufacture of carbon fibre, an improved understanding of lignin and its conversion to materials would be required (Baker and Rials 2013).

# References

Bajpai P (2013) Update on carbon fiber. Smithers Rapra, UK

Baker DA, Rials TG (2013) Recent advances in low-cost carbon fibre manufacture from lignin. J Appl Polym Sci 130:713–28

Baker DA, Harper, DP, Rials TG (2012) Extended abstract in book of abstracts of the fibre society 2012 Fall Conference, Boston Convention & Exhibition Center, Boston, MA, USA, 7–9 Nov 2012. Available at: http://www.thefibresociety.org/Assets/Past_Meetings/PastMtgs_Home.html

Black SK, Hames, BR, Myers MD (1998). U.S. Patent 5,730,837, assigned to Midwest Research Institute

Bozell JJ (2010a) Connecting biomass and petroleum processing with a chemical bridge. Science 329:522–523

Bozell JJ (2010b) Fractionation in the biorefinery. BioResources 5(3):1326–1327

Bozell JJ, O'Lenick C, Warwick S (2011) Biomass fractionation for the biorefinery: heteronuclear multiple quantum coherence–nuclear magnetic resonance investigation of lignin isolated from solvent fractionation of switchgrass. J Agric Food Chem 59:9232–9242

Chung DL (1994) Carbon fibre composites. Boston, Butterworth Heinemann, pp 3–65

Donnet JB, Bansal RC (1990) Carbon fibres, 2nd edn. Marcel Dekker, New York, pp 1–145

Frank E, Steudle LM, Ingildeev D, Spörl JM, Buchmeiser MR (2014) Carbon fibres: precursor systems, processing, structure, and properties. Angew Chem Int Ed 53:2–39

Langholtz M, Downing M, Graham R, Baker F (2014) Lignin-derived carbon fibre as a co product of refining cellulosic biomass. SAE Int J Mater Manf 7(1):115–121

Lignol Energy Corporation (2010) Lignol announces biorefining technology breakthrough with AlcellPlusTM, http://www.lignol.ca/news/AlcellPlus_Press_Release_FINAL_Oct_19_2010.pdf, Company Press Release, 19 Oct 2010

Lignol Energy Corporation (2011) Lignol develops first renewable chemical product. Company press release, 11Apr 2011

Minus ML, Kumar S (2005) The processing, properties and structure of carbon fibres. JOM 57(2), 52–58

Minus ML, Kumar S (2007) Carbon fibre. Kirk-Othmer Encycl Chem Technol 26:729–749

Morck R, Yoshida H, Kringstad KP, Hatakeyama H (1986) Fractionation of kraft lignin by successive extraction with organic solvents. Holzforschung 42:111–116

Wagemann K (2010) The German lignocellulose feedstock ... Biofuels Bioprod Biorefin 4:263–267

Watt W (1985) In: Kelly A, Rabotnov YuN (eds) Handbook of composites—volume I. Elsevier Science, Holland, pp 327–387

# Index

**A**

Accessibility, 38
Acetic acid, 38, 39, 43, 46, 48
Acetic acid pulping, 46
Acetic anhydride, 48, 56
Acetone, 39, 65
Acetylation technique, 48
Acidification, 35
Acid promoter, 40
Acrylonitrile, 50, 55
Activated carbon, 14, 44, 46, 55
Adsorption, 56
Aeronautics, 2
Aerospace, 2, 19, 20, 25, 26, 29
Agricultural, 6
Aircraft, 2, 20, 26
Airfoil, 19
Alcell lignin, 46, 47, 52
Alcohol pulping and recovery process, 39
Aldehyde group, 13
Alkali lignin, 44, 52
Alkali softwood lignin, 44
Ammonia, 20
Amorphous, 3, 4, 18
Angiosperm, 12
Animal feed, 6c
Anisotropy, 20
Annual plant, 12
Aromatic biopolymer, 3
Aromatics, 13
Automobile industry, 3, 20
Automobile manufacturer, 20
Automotive, 2, 19, 26, 31
Aviation, 19

**B**

Bagasse, 13
Bamboo, 13, 34, 51
Benzene, 14
Binder, 6
Biocompatibility, 20
Bioethanol, 38
Biofuel, 13, 14, 69
Biogas steam explosion, 38
Biomass, 3, 4, 12, 38–40, 66, 69
Biomass pre-treatment, 38
Bio-oil, 51, 54
Biopolymer, 3
Biorefinery, 30, 51, 55, 69
Birch wood, 45, 46
Bitumen, 14
Black liquor, 33, 35–37, 40, 43, 49, 50
Blend, 50, 52–54
Boat, 20
Boilers, 6
Brittleness, 50, 52, 53
BTX, 14

**C**

Cancer treatment, 2
Carbohydrates, 34, 35, 39, 49
Carbonation, 1
Carbon cracker, 14
Carbon cycle, 4, 12
Carbon dioxide, 35, 49, 69
Carbon fibre-reinforced plastics, 2
Carbon fibres, 1–3, 7, 8, 14, 17–20, 26, 27, 29, 31, 43–48, 50–56, 65, 69
Carbon nanofibres, 2, 55

© The Author(s) 2017
P. Bajpai, *Carbon Fibre from Lignin*, SpringerBriefs in Materials,
DOI 10.1007/978-981-10-4229-4

Catalyst, 46, 48
Cellulose, 3, 4, 11, 12, 33, 34, 38–40, 55
Cellulosic ethanol, 4
Cellulosic fuel ethanol, 3
Cement, 6
Cement additives, 14
Cereal straw, 13
Charcoal, 51
Chemical pulping, 3, 4, 33, 34
Chemical recovery process, 6, 34
Chemical stability, 1, 20
Coal tar, 20
Cogeneration, 6
Composite materials, 1, 2, 17, 26, 31, 54
Compression moulding, 19
Compression wood, 12
Compressive strength, 20
Concrete, 6
Condensation, 37, 49, 51
Coniferyl alcohol, 4, 12
Construction, 2, 6
Cooling system, 6
Copolymer, 50, 52, 55, 57
Corrugated packaging, 34
Covalent bond, 12
Creep resistance, 18, 20
Cross linking, 55
Crystalline, 18
Crystallinity, 40
Crystal structure, 1, 18

**D**
Defibrillation, 38
Delignification, 13, 35, 51
Demethylation, 45
Density, 1, 20
Desorption, 56
Diaryl methane type, 37
Dilute acid hydrolysis, 39
Dispersing agent, 6
Distillation, 39
DMSO, 50, 55
Dry spinning, 7, 31, 63
Dry spinning method, 44

**E**
Eelectron beam irradiation, 56
Electrical conductivity, 54
Electrical resistivity, 20
Electronic device, 3
Electronics, 2
Electrospinning, 55, 56, 65
Elemental composition, 37, 48, 51, 56
Elongation, 46, 53, 57

Emulsifying agent, 6
Energy, 3, 6, 14, 20, 29, 30, 35, 37–40, 49, 56, 69, 70
Energy efficiency, 57
Energy Refinery, 14
Environmental friendly pulping process, 38
Environmental impact, 33, 38
Enzymatic dehydrogenative polymerization, 4
Enzymatic hydrolysis, 38
Enzyme, 11
Ethanol, 4, 33, 39, 40, 43, 46, 55, 69
Evaporation, 35, 45
Extraction, 43, 46, 49
Extrusion, 7, 20, 31, 44, 47, 48, 52, 54, 56, 64, 66

**F**
Fatigue resistance, 1, 17
Fermentation, 33, 38
Fibre glass insulation, 6
Fibre morphology, 2, 65
Fibre spinning, 7, 31, 49, 51–55, 70
Fibrous mat, 56
Fishing pole, 20
Fishing rod, 2
Forestry, 6
Formaldehyde, 51, 55
Formic acid, 39
Fossil fuel, 6, 36, 69
4-O-5 aryl ether structure, 36
Fractionation, 40, 46, 52, 70
Fuel cell, 3
Fuel-efficient, 3
Fusing, 17, 52

**G**
Glass, 3, 6, 18, 49, 66
Glass transition temperature, 54, 64, 65
Graminaceous, 12
Graphite, 21
Graphite fibre, 17, 44
Graphitization, 3
Grass, 12, 39, 40
Grocery bag, 34
Guaiacyl, 12, 13
Gymnosperm, 12

**H**
Hardwood, 4–6, 12, 34, 39, 46, 47, 49–52, 56, 57, 65
Hardwood kraft lignin, 44, 46–48, 50, 53, 54
Hardwood lignin, 13, 45–47, 52–56, 65
Hardwood sawdust, 51
Heat, 39, 48, 49, 54

Heat resistance, 1
Hemicelluloses, 35, 38, 40, 43
Hexamethylenetetramine, 55
High-end automotive, 29
Higher modulus, 17
High-grade lignin, 14
Highly refractory material, 37
High modulus, 19, 20
High modulus of elasticity, 18
High specific modulus, 1
High-strength materials, 1
High tensile energy absorption, 18
High tensile modulus, 20
High thermal conductivity, 20
Hydrochloric acid, 51, 55
Hydrogenation, 45
Hydrolysis, 34, 37–39, 49
Hydrophobic material, 12
Hydrosulfide anion, 36

**I**
Industrial lignin, 64
Injection moulding, 19
Interfacial adhesion, 54
Intermediate modulus, 19
Irradiation, 45
Isotherm, 50, 56

**K**
Kappa number, 35
Kayocarbon fibre, 7, 31, 44
Kevlar fibre, 17
Kraft, 14, 33–37, 43, 47, 49, 50, 52, 56, 57, 65
Kraft cooking, 37
Kraft lignin, 7, 14, 33, 34–37, 44, 47, 49, 50, 53, 65
Kraft pulping, 33–37, 39, 49

**L**
Lightweight, 1–3, 17, 25, 26, 69
Lignifications, 11
Lignin, 3, 4, 6, 7, 11–14, 21, 30, 31, 33, 35–37, 39, 40, 43–57, 64–66, 69, 70
Lignin-carbohydrate bond, 49
Lignin monomers, 6
Lignin-poly ethylene oxide, 52
Lignin refining technique, 70
Lignocellulosic biomass, 39, 40
Lignocellulosic material, 4, 38, 39
Lignosulfonate, 33, 37, 44, 45, 50
Lignum, 11
Limb prostheses, 2

Liquid crystal projector, 3
Liquid moulding, 19
Low density, 1, 20
Low electrical resistivity, 20
Low modulus, 19
Low purity, 14
Low thermal expansion coefficient, 20

**M**
Macromolecules, 7, 13, 31
Magazine, 34
Marine, 2
Mechanical properties, 1, 17, 18, 44, 48, 50, 54, 65
Mechanical strength, 4, 12, 34, 56
Mechanical treatment, 33
Melting, 17, 46, 64
Melt spinning, 7, 17, 21, 30, 40, 44, 45, 48, 63–65, 70
Methanol, 39
Methoxyl group, 13
Micro graphite, 1
Micronutrient, 6
Milled hardwood lignin, 4
Military, 19
Modulus, 17–19, 31, 46–48, 53, 54, 57
Modulus of elasticity, 19
Molecular weight, 4, 39, 40, 44, 45, 47–49, 54, 65
Monolignol, 12
Montmorillonite, 51, 54
Morphology, 56

**N**
Nanotubes, 54
Nitric acid, 38
N-methylmorpholine-N-oxide, 55
NMR, 51
Non-woody material, 43

**O**
Oganosolv pulping, 3
Oil drilling, 3
Oil-drilling mud, 6
Oligomer, 55
Organic acids, 39
Organic amine, 48
Organic solvent, 39, 45, 64, 65
Organocell process, 39
Organoclay, 51, 54
Organosolv, 7, 14, 33, 37, 39, 40, 44, 46, 53–55, 57, 65, 70

Organosolv fractionation method, 40
Organosolv lignin, 33, 39, 52, 54
Organosolv pulping, 39
Oxalic acid, 51
Oxidation, 1, 20, 30, 49, 51, 64
Oxidative thermostabilization, 47, 48, 50, 51, 54–56

**P**
PAN-based carbon fibres, 1, 3, 7, 31
Particle board, 6
P-coumaryl alcohol, 4, 12
Peeling, 34
Pesticide, 6
Petroleum pitch, 3, 20, 30
Phenol, 12, 45, 55
Phenolated Hardwood lignin, 46
Phenol derivatives, 14
Phenolic hydroxyl group, 13, 36
Phenolic resins, 14
Phenolysis, 45
Phenyl propane, 12, 43
Pitch, 1, 18, 21, 22, 26, 27, 29, 43, 45, 50–52, 57
Plasticizer, 7, 31, 44, 48
Plywood, 6
Polyacrylonitrile (PAN), 1, 3, 18, 26, 29, 50, 51, 55
Polydispersity, 49, 52
Polyethylene glycol, 51
Polyethylene oxide, 52, 53
Polyethylene terephthalate, 53
Polymer, 4, 11, 12, 17, 20, 53–55, 63–65
Polyolefin, 30
Polyphenolic macromolecule, 3
Polypropylene, 52, 53
Polysaccharides, 4, 12
Polyvinyl alcohol, 7, 31, 44
Poplar, 40
Power, 3, 39
Precipitation, 35, 49
Pressure tank, 19
Propylene, 20
Proteins, 4
P-toluene sulphonic acid, 45
Pulping, 7, 12, 13, 33–39, 43, 46
Pulping operation, 6
Pulp mill, 6, 35–37
Pultrusion, 19
Purification, 14, 56, 65, 70
Pyridine, 48

Pyrolysis, 1, 51
Pyrolysis lignin, 51
Pyrolytic lignin, 51, 52, 54

**R**
Radiolucency, 2
Rayon, 1, 26
Rayon-based carbon fibre, 1
Rayon strand, 1
Reaction wood, 12
Recovery boilers, 35–37, 49
Refinery, 14
Refining, 70
Regenerated cellulose, 3
Regenerative medicine, 2
Reinforcement, 51
Renewable raw material, 36, 69
Renewable resource, 3
Resin, 19, 36, 55
Rocket, 2
Roving, 18
Rubber, 6

**S**
Satellite, 2
Sclereid cell, 12
Second generation bioethanol, 39
Sequestrant, 6
Shrinking, 17
Sinapyl alcohol, 4, 12
Sizing, 64
Soda pulping, 37
Sodium hydroxide, 34, 44
Sodium sulphide, 34
Softwood, 4, 6, 12, 13, 34, 36, 39, 49, 50, 52, 56, 57
Softwood kraft lignin, 44, 46, 48–50, 54, 56
Softwood lignin, 4, 5, 12, 47, 56
Solid fuel, 35
Solvent, 17, 56, 63, 64, 66
Solvent extraction, 56
Solvolysis, 51
Specialty paper, 36
Specific gravity, 17
Specific modulus, 17
Specific strength, 1, 69
Spinneret, 63, 64
Spinning, 1, 3, 7, 20, 31, 44, 45, 47–50, 52, 55, 64–66, 70
Sporting goods, 19, 25, 29
Stabilization, 3, 49

Standard modulus, 19
Steam explosion, 33, 38, 39
Steam explosion lignin, 7, 33
Steam explosion process, 38
Steel, 3, 17–20, 69
Stiffness, 2, 19, 69
Strength, 1, 2, 4, 11, 17, 18, 20, 26, 30, 31, 34, 43–48, 50, 51, 53–55, 57, 64, 66, 70
Structural alignment, 56
Structural orientation, 17
Sulfuric acid, 40, 46, 49
Sulphite process, 33
Sulphur content, 39
Surface treatment, 1
Surfactant, 6
Switch grass, 4, 40
Synthetic fibre, 69
Synthetic polymer, 29, 52, 53
Syringyl, 12, 13

**T**
Tennis racket, 2, 20
Tensile modulus, 19
Tertiary amine, 48
Thermal conductivity, 54
Thermal expansion, 1, 20
Thermal stability, 20, 47
Thermal treatment, 45, 49, 51, 56
Thermoplasticity, 7, 31
Thermoplastic polymer, 65
Thermostabilization, 17, 45–47
Thiolignin, 44
Three-dimensional macromolecule, 12
Todo Fir, 46
Toluene, 14
Tracheid, 12

Transportation, 2, 6
Tri-ethyl amine, 48

**U**
Ultrafiltration, 35, 49, 50
Ultra-high modulus, 19

**V**
Vacuum bagging, 19
Value-added, 4, 6, 30, 69
Vanillin, 14
Vascular plant, 4, 11
Vessel element, 12
Viscosity, 65

**W**
Water treatment, 6
Wet spinning, 7, 17, 21, 31, 50, 63, 64
Wheat straw, 34
White liquor, 34–36
Wind energy, 2, 26
Wind turbine, 19, 69
Wind turbine blade, 3, 26
Wood, 3, 4, 11–13, 35, 38, 43, 45, 55
Woven textile, 18

**X**
X-ray device, 2
X-ray diffraction data, 54
Xylem, 11, 12
Xylene, 14

**Y**
Yarn, 17

Printed in the United States
By Bookmasters